How the World Really Works

How the World Really Works

The Science Behind How We Got Here and Where We're Going

VACLAV SMIL

VIKING

VIKING
An imprint of Penguin Random House LLC
penguinrandomhouse.com

First published in hardcover in Great Britain by Viking, an imprint of
Penguin Random House Ltd., London, in 2022

First North American edition published by Viking, 2022

ISBN 9780593297063 (hardcover)
ISBN 9780593297070 (ebook)

Printed in the United States of America
5th Printing

Set in Bembo Book MT Std

Contents

Introduction

Why Do We Need This Book?

Every era has its claims to uniqueness, but while the experiences of the past three generations—that is, the decades since the end of the Second World War—may not have been as fundamentally transformative as those of the three generations preceding the beginning of the First World War, there has been no shortage of unprecedented events and advances. Most impressively, more people now enjoy a higher standard of living, and do so for more years and in better health, than at any time in history. Yet these beneficiaries are still a minority (only about a fifth) of the world's population, whose total count is approaching 8 billion people.

The second achievement to admire is the unprecedented expansion of our understanding of both the physical world and all forms of life. Our knowledge extends from grand generalizations about complex systems on the universal (galaxies, stars) and planetary (atmosphere, hydrosphere, biosphere) scale to processes at the level of atoms and genes: lines etched into the surface of the most powerful microprocessor are only about twice the diameter of human DNA. We have translated this understanding into a still-expanding array of machines, devices, procedures, protocols, and interventions that sustain modern civilization, and the enormity of our aggregate knowledge—and the ways we have deployed it in our service—is far beyond the comprehension of any individual mind.

You could meet real Renaissance men on Florence's Piazza Signoria in 1500—but not for too long after that. By the middle of the 18th century two French savants, Denis Diderot and Jean le Rond d'Alembert, could still gather a group of knowledgeable contributors to sum up the era's understanding in fairly exhaustive entries in their multi-volume *Encyclopédie, ou Dictionnaire raisonné des sciences, des arts et des métiers*. A few generations later the extent and the specialization of our knowledge advanced by orders of magnitude, with fundamental discoveries ranging from magnetic induction (Michael Faraday in 1831, the basis of electricity generation) to plant metabolism (Justus von Liebig, 1840, the basis of crop fertilization) to theorizing about electromagnetism (James Clerk Maxwell, 1861, the basis of all wireless communication).

In 1872, a century after the appearance of the last volume of the *Encyclopédie*, any collection of knowledge had to resort to the superficial treatment of a rapidly expanding range of topics, and, one and a half centuries later, it is impossible to sum up our understanding even within narrowly circumscribed specialties: such terms as "physics" or "biology" are fairly meaningless labels, and experts in particle physics would find it very hard to understand even the first page of a new research paper in viral immunology. Obviously, this atomization of knowledge has not made any public decision-making easier. Highly specialized branches of modern science have become so arcane that many people employed in them are forced to train until their early or mid-thirties in order to join the new priesthood.

They may share long apprenticeships, but too often they cannot agree on the best course of action. The SARS-CoV-2 pandemic made it clear that disagreements among experts may extend even to such seemingly simple decisions as wearing a face mask. By the end of March 2020 (three months into the pandemic) the World Health Organization still advised against doing so unless a person was infected, and the reversal came only in early June 2020. How can those without any special knowledge take sides or make any sense of these disputes that now often end in retractions or the dismantling of previously dominant claims?

Still, such continuing uncertainties and disputes do not excuse the extent to which most people misunderstand the fundamental workings of the modern world. After all, appreciating how wheat is grown (chapter 2) or steel is made (chapter 3) or realizing that globalization is neither new nor inevitable (chapter 4) are not the same as asking that somebody comprehend femtochemistry (the study of chemical reactions at timescales of 10^{-15} seconds, Ahmed Zewail, Nobel Prize in 1999) or polymerase chain reactions (the rapid copying of DNA, Kary Mullis, Nobel Prize in 1993).

Why then do most people in modern societies have such a superficial knowledge about how the world really works? The complexities of the modern world are an obvious explanation: people are constantly interacting with black boxes, whose relatively simple outputs require little or no comprehension of what is taking place inside the box. This is as true of such ubiquitous devices as mobile phones and laptops (typing a simple query does the trick) as it is of mass-scale procedures such as vaccination (certainly the best planetary example of 2021, with, typically, the rolling up of a sleeve being the only comprehensible part). But explanations of this comprehension deficit go beyond the fact that the sweep of our knowledge encourages specialization, whose obverse is an increasingly shallow understanding—even ignorance—of the basics.

Urbanization and mechanization have been two important reasons for this comprehension deficit. Since the year 2007, more than half of humanity has lived in cities (more than 80 percent in all affluent countries), and unlike in the industrializing cities of the 19th and early 20th centuries, jobs in modern urban areas are largely in services. Most modern urbanites are thus disconnected not only from the ways we produce our food but also from the ways we build our machines and devices, and the growing mechanization of all productive activity means that only a very small share of the global population now engages in delivering civilization's energy and the materials that comprise our modern world.

America now has only about 3 million men and women (farm owners and hired labor) directly engaged in producing food—people who actually plow the fields, sow the seeds, apply fertilizer, eradicate weeds, harvest the crops (picking fruit and vegetables is the most labor-intensive part of the process), and take care of the animals. That is less than 1 percent of the country's population, and hence it is no wonder that most Americans have no idea, or only some vague notion, about how their bread or their cuts of meat came to be. Combines harvest wheat—but do they also harvest soybeans or lentils? How long does it take for a tiny piglet to become a pork chop: weeks or years? The vast majority of Americans simply don't know—and they have plenty of company. China is the world's largest producer of steel—smelting, casting, and rolling nearly a billion tons of it every year—but all of that is done by less than 0.25 percent of China's 1.4 billion people. Only a tiny percentage of the Chinese population will ever stand close to a blast furnace, or see the continuous casting mill with its red ribbons of hot, moving steel. And this disconnect is the case across the world.

The other major reason for the poor, and declining, understanding of those fundamental processes that deliver energy (as food or as fuels) and durable materials (whether metals, non-metallic minerals, or concrete) is that they have come to be seen as old-fashioned—if not outdated—and distinctly unexciting compared to the world of information, data, and images. The proverbial best minds do not go into soil science and do not try their hand at making better cement; instead they are attracted to dealing with disembodied information, now just streams of electrons in myriads of microdevices. From lawyers and economists to code writers and money managers, their disproportionately high rewards are for work completely removed from the material realities of life on earth.

Moreover, many of these data worshippers have come to believe that these electronic flows will make those quaint old material necessities unnecessary. Fields will be displaced by urban high-rise agriculture, and synthetic products will ultimately eliminate the need to grow any food at all. Dematerialization, powered by artificial intelligence, will end our dependence on shaped masses of metals and processed minerals, and eventually we might even do without the Earth's environment: who needs it if we are going to terraform Mars? Of course, these are all not just grossly premature predictions, they are fantasies fostered by a society where fake news has become common and where reality and fiction have commingled to such an extent that gullible minds, susceptible to cult-like visions, believe what keener observers in the past would have mercilessly perceived as borderline or frank delusion.

None of the people reading this book will relocate to Mars; all of us will continue to eat staple grain crops grown in soil on large expanses of agricultural land, rather than in the skyscrapers imagined by the proponents of so-called urban agriculture; none of us will live in a dematerialized world that has no use for such irreplaceable natural services as evaporating water or pollinating plants. But delivering these existential necessities will be an increasingly challenging task, because a large share of humanity lives in conditions that the affluent minority left behind generations ago, and because the growing demand for energy and materials has been stressing the biosphere so much and so fast that we have imperiled its capability to keep its flows and stores within the boundaries compatible with its long-term functioning.

To give just a single key comparison, in 2020 the average annual per capita energy supply of about 40 percent of the world's population (3.1 billion people, which includes nearly all people in sub-Saharan Africa) was no higher than the rate achieved in both Germany and France in 1860! In order to approach the threshold of a dignified standard of living, those 3.1 billion people will need at least to double—but preferably triple—their per capita energy use, and in doing so multiply their electricity supply, boost their food production, and build essential urban, industrial, and transportation infrastructures. Inevitably, these demands will subject the biosphere to further degradation.

And how will we deal with unfolding climate change? There is now a widespread consensus that we need to do *something* to prevent many highly undesirable consequences, but what kind of action, what sort of behavioral transformation would work best? For those who ignore the energetic and material imperatives of our world, those who prefer mantras of green solutions to understanding how we have come to this point, the prescription is easy: just decarbonize—switch from burning fossil carbon to converting inexhaustible flows of renewable energies. The real wrench in the works: we are a fossil-fueled civilization whose technical and scientific advances, quality of life, and prosperity rest on the combustion of huge quantities of fossil carbon, and we cannot simply walk away from this critical determinant of our fortunes in a few decades, never mind years.

Complete decarbonization of the global economy by 2050 is now conceivable only at the cost of unthinkable global economic retreat, or as a result of extraordinarily rapid transformations relying on near-miraculous technical advances. But who is going, willingly, to engineer the former while we are still lacking any convincing, practical, affordable global strategy and technical means to pursue the latter? What will actually happen? The gap between wishful thinking and reality is vast, but in a democratic society no contest of ideas and proposals can proceed in rational ways without all sides sharing at least a modicum of relevant information about the real world, rather than trotting out their biases and advancing claims disconnected from physical possibilities.

This book is an attempt to reduce the comprehension deficit, to explain some of the most fundamental ruling realities governing our survival and our prosperity. My goal is not to forecast, not to outline either stunning or depressing scenarios of what is to come. There is no need to extend this popular—but consistently failing—genre: in the long run, there are too many unexpected developments and too many complex interactions that no individual or collective effort can anticipate. Nor will I advocate any specific (biased) interpretations of reality, either as a source of despair or of boundless expectations. I am neither a pessimist nor an optimist; I am a scientist trying to explain how the world really works, and I will use that understanding in order to make us better realize our future limits and opportunities.

Inevitably, this kind of inquiry must be selective, but every one of the seven key topics chosen for closer examination passes the muster of existential necessity: there are no frivolous choices in the lineup. The first chapter of this book shows how our high-energy societies have been steadily increasing their dependence on fossil fuels in general and on electricity, the most flexible form of energy, in particular. Appreciation of these realities serves as a much-needed corrective to the now-common claims (based on a poor understanding of complex realities) that we can decarbonize the global energy supply in a hurry, and that it will take only two or three decades before we rely solely on renewable energy conversions. While we are converting increasing shares of electricity generation to new renewables (solar and wind, as opposed to the long-established hydroelectricity) and putting more electric cars on the roads, decarbonizing trucking, flying, and shipping will be a much greater challenge, as will the production of key materials without relying on fossil fuels.

The second chapter of this book is about the most basic survival necessity: producing our food. Its focus is on explaining how much of what we rely on to survive, from wheat to tomatoes to shrimp, has one thing in common: it requires substantial, direct and indirect, fossil fuel inputs. Awareness of this fundamental dependence on fossil fuels leads to a realistic understanding of our continued need for fossil carbon: it is relatively easy to generate electricity by wind turbines or solar cells rather than by burning coal or natural gas—but it would be much more difficult to run all field machinery without liquid fossil fuels and to produce all fertilizers and other agrochemicals without natural gas and oil. In short, for decades it will be impossible to adequately feed the planet without using fossil fuels as sources of energy and raw materials.

The third chapter explains how and why our societies are sustained by materials created by human ingenuity, focusing on what I call the four pillars of modern civilization: ammonia, steel, concrete, and plastics. Understanding these realities exposes the misleading nature of recently fashionable claims about the dematerialization of modern economies dominated by services and miniaturized electronic devices. The relative decline of material needs per unit of many finished products has been one of the defining trends of modern industrial developments. But in absolute terms, material demands have been rising even in the world's most affluent societies, and they remain far below any conceivable saturation levels in low-income countries where the ownership of well-built apartments, kitchen appliances, and air conditioning (to say nothing about cars) remains a dream for billions of people.

The fourth chapter is the story of globalization, or how the world has become so interconnected by transportation and communication. This historical perspective shows how old (or indeed ancient) the origins of this process are, and how recent is its highest—and finally truly global—extent. And a closer look makes it clear that there is nothing inevitable about the future course of this ambivalently perceived (much praised, much questioned, and much criticized) phenomenon. Recently, there have been some clear retreats around the world, and a general trend toward populism and nationalism, but it is not clear how far these will continue, or to what extent these changes will be modified due to a combination of economic, security, and political considerations.

The fifth chapter provides a realistic framework for judging the risks we face: modern societies have succeeded in eliminating or reducing many previously mortal or crippling risks—polio and giving birth, for example—but many perils will always be with us, and we repeatedly fail to make proper risk assessments, both underestimating and exaggerating the dangers we face. After finishing this chapter, readers will have a good appreciation of the relative risks of many common involuntary exposures and voluntary activities (from falling at home to flying between continents; from living in a hurricane-prone city to parachuting)—and, cutting through the diet industry nonsense, we will see a range of options of what we could eat to help us live longer.

The sixth chapter will look first at how unfolding environmental changes might affect our three existential necessities: oxygen, water, and food. The rest of the chapter will focus on global warming, the change that has dominated recent environmental concerns and has led to the emergence of new—near apocalyptic—catastrophism on one hand, and complete denials of the process on the other. Instead of recounting and adjudging these contested claims (too many books have already done so), I will stress that, contrary to widespread perceptions, this is not a recently discovered phenomenon: we have understood the fundamentals of this process for more than 150 years.

Moreover, we have been aware of the actual degree of warming associated with the doubling of atmospheric CO_2 for more than a century, and we were warned about the unprecedented (and unrepeatable) nature of this planetary experiment more than half a century ago (uninterrupted, accurate measurements of CO_2 began in 1958). But we have chosen to ignore these explanations, warnings and recorded facts. Instead, we have multiplied our reliance on the combustion of fossil fuels, resulting in a dependence that will not be severed easily, or inexpensively. How rapidly we can change this remains unclear. Add to this all other environmental worries, and you must conclude that the key existential question—can humanity realize its aspirations within the safe boundaries of our biosphere?—has no easy answers. But it is imperative that we understand the facts of the matter. Only then can we tackle the problem effectively.

In the closing chapter I will look to the future, specifically at the recent opposing propensities to embrace catastrophism (those that say there are just years left before the final curtain descends on modern civilization) and techno-optimism (those that predict that the powers of invention will open unlimited horizons beyond the confines of the Earth, turning all terrestrial challenges into inconsequential histories). Predictably, I have little use for either of these positions, and my perspective will find no favor with either doctrine. I do not foresee any imminent break with history in either direction; I do not see any already predetermined outcomes, but rather a complicated trajectory contingent on our—far from foreclosed—choices.

This book rests on two foundations: abundant scientific findings and half a century of my research and book-writing. The first includes items ranging from such classic contributions as the pioneering elucidations of energy conversions and of the greenhouse gas effect from the 19th century, through to the very latest assessments of global challenges and risk probabilities. And this far-reaching book could not have been written without my decades of interdisciplinary studies distilled in my many other books. Rather than resorting to an ancient comparison of foxes and hedgehogs (a fox knows many things, but a hedgehog knows one big thing), I tend to think about modern scientists as either the drillers of ever-deeper holes (now the dominant route to fame) or scanners of wide horizons (now a much-diminished group).

Drilling the deepest possible hole and being an unsurpassed master of a tiny sliver of the sky visible from its bottom has never appealed to me. I have always preferred to scan as far and as wide as my limited capabilities have allowed me to do. My main area of interest throughout my life has been energy studies, because a satisfactory grasp of that vast field requires you to combine an understanding of physics, chemistry, biology, geology, and engineering with an attention to history and to social, economic, and political factors.

Nearly half of my now more than 40 (and mainly more academic) books deal with various aspects of energy, from wide-ranging surveys of general energetics and energy throughout history to closer looks at individual fuel categories (oil, natural gas, biomass) and specific properties and processes (power density, energy transitions). The rest of my output betrays my interdisciplinary quests: I have written about such fundamental phenomena as growth—in all of its natural and anthropogenic guises—and risk; about the global environment (the biosphere, biogeochemical cycles, global ecology, photosynthetic productivity, and harvests), food and agriculture, materials (above all, steel and fertilizers), technical advances, and the progress and retreat of manufacturing, and also about ancient Roman and modern American history and Japanese food.

Inevitably, this book—the product of my life's work, and written for the layperson—is a continuation of my long-lasting quest to understand the basic realities of the biosphere, history, and the world we have created. And it also does, yet again, what I have been steadfastly doing for decades: it strongly advocates for moving away from extreme views. Recent (and increasingly strident or increasingly giddy) advocates of such positions will be disappointed: this is not the place to find either laments about the world ending in 2030 or an infatuation with astonishingly transformative powers of artificial intelligence arriving sooner than we think. Instead, this book tries to provide a foundation for a more measured and necessarily agnostic perspective. I hope that my rational, matter-of-fact approach will help readers to understand how the world really works, and what our chances are of seeing it offer better prospects to the coming generations.

But before you plunge into the specific topics, I have a warning as well as a possible request. This book teems with numbers (all metric) because the realities of the modern world cannot be understood only by qualitative descriptions. Many numbers in this book are, inevitably, either very large or very small, and such realities are best treated in terms of orders of magnitude, labelled with globally valid prefixes. Should you not have a grounding in these matters, the appendix on understanding numbers, large and small, takes care of that, and hence some readers might find it profitable to begin this book from its end. Otherwise, I'll see you in chapter 1 for a closer, quantitative look at energies. It's a perspective that should never go out of fashion.

It is really fun to be a complex system thinker. Think about sitting with a group of 30~ or 40~ friends who all like to brag about their smartness, do you want to be the shiny star?

1. Understanding Energy

Fuels and Electricity

Consider a benign science fiction scenario: not travel to distant planets in search of life, but the Earth and its inhabitants as targets of remote monitoring by an exceedingly sapient civilization that sends its probes to nearby galaxies. Why do they do this? Just for the satisfaction of systematic understanding, and perhaps also to avoid dangerous surprises should the third planet orbiting around an unremarkable star in a spiral galaxy become a threat, or perhaps in case they should require a second home. Hence this planet keeps periodic tabs on Earth.

Let us imagine that a probe approaches our planet once every 100 years and that it is programmed to make a second pass (a closer inspection) only when it detects a previously unobserved kind of energy conversion—the changing of energy from one form to another—or a new physical manifestation dependent on it. In fundamental physical terms, any process—be it rain, a volcanic eruption, plant growth, animal predation, or the growth of human sapience—can be defined as a sequence of energy conversions, and for a few hundred million years after the Earth's formation the probes would see only the same varied, but ultimately monotonous, displays of volcanic eruptions, earthquakes and atmospheric storms.

Fundamental shifts

The first microorganisms emerge nearly 4 billion years ago but passing probes do not register them, as these life forms are rare and remain hidden, associated with alkaline hydrothermal vents at the ocean's floor. The first occasion for a closer look arises as early as 3.5 billion years ago, when a passing probe records the first simple, single-celled photosynthetic microbes in shallow seas: they absorb near-infrared radiation—that which is just beyond the visible spectrum—and do not produce oxygen.[1] Hundreds of millions of years then elapse with no signs of change before cyanobacteria begin to use the energy of the visible incoming solar radiation to convert CO_2 and water into new organic compounds and release oxygen.[2]

This is a radical shift that will create Earth's oxygenated atmosphere, yet a long time elapses before new, more complex aquatic organisms are seen 1.2 billion years ago, when the probes document the rise and diffusion of brilliantly colored red algae (due to the photosynthetic pigment phycoerythrin) and of much larger, brown algae. Green algae arrive nearly half a billion years later, and because of the new proliferation of marine plants the probes get better sensors to monitor the sea floor. This pays off, as more than 600 million years ago the probes make another epochal discovery: the existence of the first organisms made of differentiated cells. These flattish, soft, bottom-dwelling creatures (known as Ediacaran fauna after their Australian domicile) are the first simple animals requiring oxygen for their metabolism and, unlike algae that are merely tossed by waves and currents, they are mobile.[3]

And then the probes begin to document what are, comparatively speaking, rapid changes: instead of passing over lifeless continents and waiting hundreds of millions of years before logging another epochal shift, they begin to record the rising, cresting, and subsiding waves of the emergence, diffusion, and extinction of a huge variety of species. This period starts with the Cambrian explosion of small marine bottom-dwellers (541 million years ago, dominated first by trilobites) through the arrival of the first fishes, amphibians, land plants, and four-legged (and hence exceptionally mobile) animals. Periodic extinctions reduce, or sometimes almost eliminate, this variety, and even just 6 million years ago the probes do not find any organism dominating the planet.[4] Not long afterwards, the probes nearly miss the significance of a mechanical shift with enormous energetic implications: many four-legged animals briefly stand or awkwardly walk on two legs, and more than 4 million years ago this form of locomotion becomes the norm for small ape-like creatures that begin spending more time on land than in trees.[5]

Now the intervals between reporting something noteworthy to their home base shrink from hundreds of millions to mere hundreds of thousands of years. Eventually the descendants of these early bipeds (we classify them as hominins, belonging to the genus *Homo*, along the long line of our ancestors) do something that puts them on an accelerated path to planetary dominance. Several hundred thousand years ago, the probes detect the first extrasomatic use of energy—external to one's body; that is, any energy conversion besides digesting food—when some of these upright walkers master fire and begin to use it deliberately for cooking, comfort, and safety.[6] This controlled combustion converts the chemical energy of plants into thermal energy and light, enabling the hominins to eat previously hard-to-digest foods, warming them through the cold nights, and keeping away dangerous animals.[7] These are the first steps toward deliberately shaping and controlling the environment on an unprecedented scale.

This trend intensifies with the next notable change, the adoption of crop cultivation. About 10 millennia ago, the probes record the first patches of deliberately cultivated plants as a small share of the Earth's total photosynthesis becomes controlled and manipulated by humans who domesticate—select, plant, tend, and harvest—crops for their (delayed) benefit.[8] The first domestication of animals soon follows. Before that happens, human muscles are the only prime movers—that is, converters of chemical (food) energy to the kinetic (mechanical) energy of labor. Domestication of working animals, starting with cattle some 9,000 years ago, supplies the first extrasomatic energy other than that of human muscles—they are used for field work, for lifting water from wells, for pulling or carrying loads, and for providing personal transportation.[9] And much later come the first inanimate prime movers: sails, more than five millennia ago; waterwheels, more than two millennia ago; and windmills, more than a thousand years ago.[10]

Afterwards, the probes don't have much to observe, following the arrival of another period of (relative) slowdown: century after century, there is just repetition, stagnation, or the slow growth and diffusion of these long-established conversions. In the Americas and in Australia (lacking any draft animals and any simple mechanical prime movers), all work before the arrival of Europeans is done by human muscles. In some of the Old World's preindustrial regions, harnessed animals, wind and running or falling water energize significant shares of grain milling, oil pressing, grinding, and forging, and draft animals become indispensable for heavy field work (plowing above all, as harvesting is still done manually), transporting goods, and waging wars.

But at this point, even in societies with domesticated animals and mechanical prime movers, much of the work is still done by people. My estimate, using necessarily approximate past totals of working animals and people and assuming typical daily work rates based on modern measurements of physical exertion, is that—be it at the beginning of the second millennium of the Common Era or 500 years later (in 1500, at the beginning of the early modern era)—more than 90 percent of all useful mechanical energy was provided by animate power, roughly split between people and animals, while all thermal energy came from the combustion of plant fuels (mostly wood and charcoal, but also straw and dried dung).

And then in 1600 the alien probe will spring into action, and spot something unprecedented. Rather than relying solely on wood, an island society is increasingly burning coal, a fuel produced by photosynthesis tens or hundreds of millions of years ago and fossilized by heat and pressure during its long underground storage. The best reconstructions show that coal as a heat source in England surpasses the use of biomass fuels around 1620 (perhaps even earlier); by 1650 the burning of fossil carbon supplies two-thirds of all heat; and the share reaches 75 percent by 1700.[11] England has an exceptionally early start: all the coalfields that make the UK the world's leading 19th-century economy are already producing coal before 1640.[12] And then, at the very beginning of the 18th century, some English mines begin to rely on steam engines, the first inanimate prime movers powered by the combustion of fossil fuel.

These early engines are so inefficient that they can be deployed only in mines where the fuel supply is readily available and does not require any transportation.[13] But for generations the UK remains the most interesting nation to the alien probe because it is an exceptional early adopter. Even by 1800, the combined coal extraction in a few European countries and the United States is a small fraction of British production.

By 1800 a passing probe will record that, across the planet, plant fuels still supply more than 98 percent of all heat and light used by the dominant bipeds, and that human and animal muscles still provide more than 90 percent of all mechanical energy needed in farming, construction, and manufacturing. In the UK, where James Watt introduced an improved steam engine during the 1770s, the Boulton & Watt company begin to build engines whose average power is equal to that of 25 strong horses, but by 1800 they have sold less than 500 of these machines, merely denting the total power provided by harnessed horses and hard-working laborers.[14]

Even by 1850, rising coal extraction in Europe and North America supplies no more than 7 percent of all fuel energy, nearly half of all useful kinetic energy comes from draft animals, about 40 percent from human muscles, and just 15 percent from the three inanimate prime movers: waterwheels, windmills, and the slowly spreading steam engines. The world of 1850 is much more akin to the world of 1700 or even of 1600 than that of the year 2000.

But by 1900 the global share of both fossil and renewable fuels and of prime movers shifts considerably as modern energy sources (coal and some crude oil) provide half of all primary energy, and traditional fuels (wood, charcoal, straw) the other half. Water turbines in hydro stations generate the first primary electricity during the 1880s; later comes geothermal electricity, and after the Second World War nuclear, solar, and wind electricity (the new renewables). But by 2020 more than half of the world's electricity will still be generated by the combustion of fossil fuels, mainly coal and natural gas.

By 1900, inanimate prime movers supply about half of all mechanical energy: coal-fired steam engines make the greatest contribution, followed by better-designed waterwheels and new water turbines (first introduced during the 1830s), windmills and brand-new steam turbines (since the late 1880s), and internal combustion engines (gasoline-fueled, also first introduced in the 1880s).[15]

By 1950, fossil fuels supply nearly three-quarters of primary energy (still dominated by coal), and inanimate prime movers—now with gasoline- and diesel-fueled internal combustion engines in the lead—provide more than 80 percent of all mechanical energy. And by the year 2000 only poor people in low-income countries depend on biomass fuels, with wood and straw providing only about 12 percent of the world's primary energy. Animate prime movers hold only a 5 percent share of mechanical energy, as human exertions and the work of draft animals are almost completely displaced by machines fueled by liquids or by electric motors.

During the past two centuries, the alien probes will have witnessed a rapid global substitution of primary energy sources, accompanied by the expansion and diversification of fossil energy supply, and the no less rapid introduction, adoption, and growth in capacity of new inanimate prime movers—first coal-fired steam engines, then internal combustion engines (piston and turbines). The most recent visit would see a truly global society built and defined by mass-scale, stationary, and mobile conversions of fossil carbon, deployed everywhere but in some of the planet's uninhabited regions.

Modern energy uses

What difference has this mobilization of extrasomatic energies made? Global primary energy supply usually refers to total (gross) production, but it is more revealing to look at energy that is actually available for conversion into useful forms. To do this, we need to subtract preconsumption losses (during coal sorting and cleaning, crude oil refining, and natural gas processing), non-energy use (mainly as feedstocks for chemical industries, and also as lubricating oils for machines ranging from pumps to aircraft turbines and as paving materials), and losses during electricity transmission. With these adjustments—and rounding heavily to avoid impressions of unwarranted accuracy—my calculations show a 60-fold increase in the use of fossil fuels during the 19th century, a 16-fold gain during the 20th century, and about a 1,500-fold increase over the past 220 years.[16]

This increasing dependence on fossil fuels is the most important factor in explaining the advances of modern civilization—and also our underlying concerns about the vulnerability of their supply and the environmental impacts of their combustion. In reality, the energy gain was substantially higher than the 1,500-fold I just mentioned, because we must take into account the concurrent increase in average conversion efficiencies.[17] In 1800, coal combustion in stoves and boilers to produce heat and hot water was no more than 25–30 percent efficient, and only 2 percent of coal consumed by steam engines was converted into useful work, resulting in an overall conversion efficiency of no higher than 15 percent. A century later, better stoves, boilers, and engines raised the overall efficiency to nearly 20 percent, and by the year 2000 the mean conversion rate was about 50 percent. Consequently, the 20th century saw a nearly 40-fold gain in useful energy; since 1800 the gain was about 3,500-fold.

To get an even clearer picture of the magnitude of these changes, we should express these rates in per capita terms. The global population rose from 1 billion in 1800 to 1.6 billion in 1900 and 6.1 billion in the year 2000, and hence the supply of useful energy rose (all values in gigajoules per capita) from 0.05 in 1800 to 2.7 in 1900 and to about 28 in the year 2000. China's post-2000 rise on the world stage was the main reason for a further increase in the global rate to about 34 GJ/capita by 2020. An average inhabitant of the Earth nowadays has at their disposal nearly 700 times more useful energy than their ancestors had at the beginning of the 19th century.

Moreover, within a lifetime of people born just after the Second World War the rate had more than tripled, from about 10 to 34 GJ/capita between 1950 and 2020. Translating the last rate into more readily imaginable equivalents, it is as if an average Earthling has every year at their personal disposal about 800 kilograms (0.8 tons, or nearly six barrels) of crude oil, or about 1.5 tons of good bituminous coal. And when put in terms of physical labor, it is as if 60 adults would be working non-stop, day and night, for each average person; and for the inhabitants of affluent countries this equivalent of steadily laboring adults would be, depending on the specific country, mostly between 200 and 240. On average, humans now have unprecedented amounts of energy at their disposal.

The consequences of this in terms of human exertion, hours of physical labor, time for leisure, and the overall standard of living are obvious. An abundance of useful energy underlies and explains all the gains—from better eating to mass-scale travel; from mechanization of production and transport to instant personal electronic communication—that have become norms rather than exceptions in all affluent countries. Recent changes on a national scale range widely: as expected, they are lower for those high-income countries whose per capita energy use was already relatively high a century ago, with a greater increase in nations that have seen the most rapid modernization of their economies since 1950, most notably Japan, South Korea, and China. Between 1950 and 2020 the United States roughly doubled the per capita useful energy provided by fossil fuels and primary electricity (to about 150 gigajoules); in Japan the rate had more than quintupled (to nearly 80 GJ/capita), and China saw an astounding, more than 120-fold, increase (to nearly 50 GJ/capita).[18]

Tracing the trajectory of useful energy deployment is so revealing because energy is not just another component in the complex structures of the biosphere, human societies, and their economies, nor just another variable in intricate equations determining the evolution of these interacting systems. Energy conversions are the very basis of life and evolution. Modern history can be seen as an unusually rapid sequence of transitions to new energy sources, and the modern world is the cumulative result of their conversions.

Physicists were the first to recognize the fundamental importance of energy in human affairs. In 1886, Ludwig Boltzmann, one of the founders of thermodynamics, spoke about free energy—energy available for conversions—as the *Kampfobjekt* (the object of struggle) for life, which is ultimately dependent on incoming solar radiation.[19] Erwin Schrödinger, winner of the Nobel Prize in Physics in 1933, summed up the basis of life: "What an organism feeds upon is negative entropy" (negative entropy or negentropy = free energy).[20] During the 1920s, following this fundamental insight of 19th- and early 20th-century physicists, the American mathematician and statistician Alfred Lotka concluded that those organisms that best capture the available energy hold the evolutionary advantage.[21]

In the early 1970s, American ecologist Howard Odum explained how "all progress is due to special power subsidies, and progress evaporates whenever and wherever they are removed."[22] And, more recently, physicist Robert Ayres has repeatedly stressed in his writings the central notion of energy in all economies: "the economic system is essentially a system for extracting, processing and transforming energy as resources into energy embodied in products and services."[23] Simply put, energy is the only truly universal currency, and nothing (from galactic rotations to ephemeral insect lives) can take place without its transformations.[24]

Given all of these readily verifiable realities, it is hard to understand why modern economics, that body of explanations and precepts whose practitioners exercise more influence on public policy than any other experts, has largely ignored energy. As Ayres noted, economics does not only lack any systematic awareness of energy's importance for the physical process of production, but it assumes "that energy doesn't matter (much) because the cost share of energy in the economy is so small that it can be ignored . . . as if output could be produced by labor and capital alone—or as if energy is merely a form of man-made capital that can be produced (as opposed to extracted) by labor and capital."[25]

Modern economists do not get their rewards and awards for being preoccupied with energy, and modern societies become concerned about it only when the supply of any main commercial form of energy appears to be threatened and its prices soar. Google's Ngram Viewer, a tool that allows you to see the popularity of terms that appeared in printed sources between 1500 and 2019, illustrates this point: during the 20th century the frequency of the term "energy price" remained quite negligible, until a sudden spike that began in the early 1970s (caused by OPEC's quintupling of crude oil prices; details of which follow later in this chapter) and peaked in the early 1980s. Once prices fell, a similarly steep decline followed, and by 2019 the term "energy price" was mentioned no more frequently than it was in 1972.

Understanding how the world really works cannot be done without at least a modicum of energy literacy. In this chapter I will first explain that energy may not be easy to define but that it is easy not to make the commonly encountered error of conflating it with power. We'll see how different forms of energy (with their specific advantages and drawbacks) and different energy densities (energy stored per unit of mass or volume, critical for energy storage and portability) have affected different stages of economic development, and I'll offer some realistic appraisals of the challenges faced by the unfolding transition to societies relying less and less on fossil carbon. As we'll see, our civilization is so deeply reliant on fossil fuels that the next transition will take much longer than most people think.

What is energy?

How do we define this fundamental quantity? The Greek etymology is clear. Aristotle, writing in his *Metaphysics*, combined ἐν (in) with ἔργον (work) and concluded that every object is maintained by ἐνέργεια.[26] This understanding endowed all objects with the potential for action, motion, and change—not a bad characterization of a potential to be transformed into other forms, be it by lifting, throwing, or burning.

Little changed over the following two millennia. Eventually, Isaac Newton (1643–1727) laid down fundamental physical laws involving mass, force, and momentum, and his second law of motion made it possible to derive the basic energy units. Using modern scientific units, 1 joule is the force of 1 newton—that is, the mass of 1 kilogram accelerated by 1 m/s^2 acting over a distance of 1 meter.[27] But this definition refers only to kinetic (mechanical) energy, and it certainly does not provide an intuitive understanding of energy in all of its forms.

Our practical understanding of energy was greatly expanded during the 19th century thanks to the era's proliferating experiments with combustion, heat, radiation, and motion.[28] This led to what is still the most common definition of energy: "the capacity for doing work"—a definition valid only when the term "work" means not only some invested labor but, as one of the leading physicists of the era put it, a generalized physical "act of producing a change of configuration in a system in opposition to a force which resists that change."[29] But that, too, is still too Newtonian to be intuitive.

There is no better way to answer the question "what is energy?" than by referring to one of the most insightful physicists of the 20th century—to the protean mind of Richard Feynman, who (in his famous *Lectures on Physics*) tackled the challenge in his straightforward manner, stressing that "energy has a large number of *different forms*, and there is a formula for each one. These are: gravitational energy, kinetic energy, heat energy, elastic energy, electrical energy, chemical energy, radiant energy, nuclear energy, mass energy."

And then comes this disarming but indubitable conclusion:

> It is important to realize that in physics today, we have no knowledge of what energy *is*. We do not have a picture that energy comes in little blobs of a definite amount. It is not that way. However, there are formulas for calculating some numerical quantity, and when we add it all together it gives . . . always the same number. It is an abstract thing in that it does not tell us the mechanism or the *reasons* for the various formulas.[30]

And so it has been. We can use formulas to calculate, very accurately, the kinetic energy of a moving arrow or of a cruising jetliner, or the potential energy of a massive boulder that is just about to tumble down from a mountain, or the thermal energy released by a chemical reaction, or the light (radiant) energy of a flickering candle or a pointed laser—but we cannot reduce these energies into a single, easily described entity in our mind.

But the slippery nature of energy has not troubled the armies of instant experts: ever since the early 1970s, when energy became a major topic of public discourse, they have opined on energy matters with ignorance and zeal. Energy is among the most elusive and most misunderstood concepts, and a poor grasp of basic realities has led to many illusions and delusions. As we have seen, energy exists in various forms, and to make it useful to us we need to convert one form of it into another type. But treating this multifaceted abstract as a monolith has been the norm, as if different forms of energy were effortlessly substitutable.

Some of these substitutions are both relatively easy and beneficial.

Replacing candles (the chemical energy of wax transformed into radiant energy) with electric lights powered by electricity generated by steam turbines (the chemical energy of fuels transformed first into heat and then into electric energy, which is then transformed into radiant energy) resulted in many obvious benefits (a safer, brighter, cheaper, more reliable kind of energy). Replacing steam- and diesel-powered railroad engines with electric drive has allowed less expensive, cleaner, and faster transportation: all sleek, high-speed trains are electric. But many desirable substitutions remain more expensive, or possible but realistically unaffordable for some time to come, or impossible at required scales—no matter how loudly their promoters extol their virtues.

Electric cars are a common example of the first category: now readily available, the best models are quite reliable, but in 2020 they were still more expensive than similarly sized vehicles powered by internal combustion engines. In terms of the second category, as I will detail in the next chapter, synthesis of the ammonia needed to produce nitrogenous fertilizers now depends heavily on natural gas as the source of hydrogen. Hydrogen could be produced by the decomposition (electrolysis) of water instead, but this route remains nearly five times as expensive as when the element is derived from abundant and inexpensive methane—and we have yet to create a mass-scale hydrogen industry. And long-distance electricity-powered commercial flight (equivalent to a kerosene-powered Boeing 787 from New York to Tokyo) is the outstanding example of the last category: as we will see, this is an energy conversion that will remain unrealistic for a long time to come.

The first law of thermodynamics states that no energy is ever lost during conversions: be that chemical to chemical when digesting food; chemical to mechanical when moving muscles; chemical to thermal when burning natural gas; thermal to mechanical when rotating a turbine; mechanical to electrical in a generator; or electrical to electromagnetic as light illuminates the page you are reading. However, all energy conversions eventually result in dissipated low-temperature heat: no energy has been *lost*, but its utility, its ability to perform useful work, is gone (the second law of thermodynamics).[31]

All forms of energy can be measured in the same units—joule is the scientific unit; calories are often used in nutritional studies. In the next chapter, when I detail the massive energy subsidies going into modern food production, we will encounter the truly existential reality of different energy qualities. Producing chicken requires energies whose total is several times higher than the energy content of the edible meat. Although we can calculate the subsidy ratio in terms of energy quantities (joules in/joules out)—there is, obviously, a fundamental difference between inputs and outputs: we cannot digest diesel oil or electricity, while lean chicken meat is an almost perfectly digestible foodstuff containing high-quality protein, an indispensable macronutrient that cannot be replaced by an equal amount of energy from lipids or carbohydrates.

There are many choices available when it comes to energy conversions, some far better than others. The high densities of chemical energy in kerosene and diesel fuel are great for intercontinental flying and shipping, but if you want your submarine to stay submerged while crossing the Pacific Ocean then the best choice is to fission enriched uranium in a small reactor in order to produce electricity.[32] And back on land, large nuclear reactors are the most reliable producers of electricity: some of them now generate it 90–95 percent of the time, compared to about 45 percent for the best offshore wind turbines and 25 percent for photovoltaic cells in even the sunniest of climates—while Germany's solar panels produce electricity only about 12 percent of the time.[33]

This is simple physics or electrical engineering, but it is remarkable how often these realities are ignored. Another common mistake is to confuse energy with power, and this is done even more frequently. It betrays an ignorance of basic physics, and one that, regrettably, is not limited to lay usage. Energy is a scalar, which in physics is a quantity described only by its magnitude; volume, mass, density, time, and speed are other ubiquitous scalars. Power measures energy per unit of time and hence it is a rate (in physics, a rate measures change, commonly per time). Establishments that generate electricity are commonly called power plants—but power is simply the rate of energy production or energy use. Power equals energy divided by time: in scientific units, it is watts = joules/seconds. Energy equals power multiplied by time: joules = watts × seconds. If you light a small votive candle in a Roman church, it might burn for 15 hours, converting the chemical energy of wax to heat (thermal energy) and light (electromagnetic energy) with an average power of nearly 40 watts.[34]

Unfortunately, even engineering publications often write about a "power station generating 1,000 MW of electricity," but that is impossible. A generating station may have installed (rated) power of 1,000 megawatts—that is, it can produce electricity at that rate—but when doing so it would generate 1,000 megawatt-hours or (in basic scientific units) 3.6 trillion joules in an hour (1,000,000,000 watts × 3,600 seconds). Analogically, an adult man's basal metabolic rate (the energy required at complete rest to run the body's essential functions) is about 80 watts, or 80 joules per second; lying prone all day a 70-kilogram man would still need about 7 megajoules (80 × 24 × 3,600) of food energy, or about 1,650 kilocalories, to maintain his body temperature, energize his beating heart, and run myriad enzymatic reactions.[35]

Most recently, a poor understanding of energy has the proponents of a new green world naively calling for a near-instant shift from abominable, polluting, and finite fossil fuels to superior, green and ever-renewable solar electricity. But liquid hydrocarbons refined from crude oil (gasoline, aviation kerosene, diesel fuel, residual heavy oil) have the highest energy densities of all commonly available fuels, and hence they are eminently suitable for energizing all modes of transportation. Here is a density ladder (all rates in gigajoules per ton): air-dried wood, 16; bituminous coal (depending on quality), 24–30; kerosene and diesel fuels, about 46. In volume terms (all rates in gigajoules per cubic meter), energy densities are only about 10 for wood, 26 for good coal, 38 for kerosene. Natural gas (methane) contains only 35 MJ/m^3—or less than 1/1,000 of kerosene's density.[36]

The implications of energy density—as well as of fuel's physical properties—for transport are obvious. Ocean liners powered by steam engines did not burn wood because, everything else being equal, firewood would have taken up 2.5 times the volume of the good bituminous coal required for a transatlantic crossing (and be at least 50 percent heavier), greatly reducing the ship's capacity to transport people and goods. There could be no natural gas–powered flight, as the energy density of methane is three orders of magnitude lower than that of aviation kerosene, and also no coal-powered flight—the density difference is not that large, but coal would not flow from wing tanks to engines.

And the advantages of liquid fuels go far beyond high energy density. Unlike coal, crude oil is much easier to produce (no need to send miners underground or scar landscapes with large open pits), store (in tanks or underground—because of oil's much higher energy density, any enclosed space can typically store 75 percent more energy as a liquid fuel than as coal), and distribute (intercontinentally by tankers and by pipelines, the safest mode of long-distance mass transfer), and hence it is readily available on demand.[37] Crude oil needs refining to separate the complex mixture of hydrocarbons into specific fuels—gasoline being the lightest; residual fuel oil the heaviest—but this process yields more valuable fuels for specific uses, and it also produces indispensable non-fuel products such as lubricants.

Lubricants are needed to minimize friction in everything from the massive turbofan engines in wide-body jetliners to miniature bearings.[38] Globally, the automotive sector, now with more than 1.4 billion vehicles on the road, is the largest consumer, followed by use in industry—with the largest markets being textiles, energy, chemicals, and food processing—and in ocean-going vessels. Annual use of these compounds now surpasses 120 megatons (for comparison, global output of all edible oils, from olive to soybean, is now about 200 megatons a year), and because the available alternatives—synthetic lubricants made from simpler, but still often oil-based, compounds rather than those derived directly from crude oil—are more expensive, this demand will grow further as these industries expand around the world.

Another product derived from crude oil is asphalt. Global output of this black and sticky material is now on the order of 100 megatons, with 85 percent of it going to paving (hot and warm asphalt mixes) and most of the rest to roofing.[39] And hydrocarbons have yet another indispensable non-fuel use: as feedstocks for many different chemical syntheses (dominated by ethane, propane, and butane from natural gas liquids) producing a variety of synthetic fibers, resins, adhesives, dyes, paints and coatings, detergents, and pesticides, all vital in myriad ways to our modern world.[40] Given these advantages and benefits, it was predictable—indeed unavoidable—that our dependence on crude oil would grow once the product became more affordable and once it could be reliably delivered on a global scale.

The shift from coal to crude oil took generations to accomplish. Commercial crude oil extraction began during the 1850s in Russia, Canada, and the US. The wells, drilled using the ancient percussion method involving the raising and dropping of a heavy cutting bit, were shallow, their daily productivities were low, and kerosene for lamps (which displaced whale oil and candles) was the main product of the simple refining of crude oil.[41] New markets for refined oil products were created only with the widespread adoption of internal combustion engines: first the gasoline-fueled (Otto cycle) machines for cars, buses, and trucks; then Rudolf Diesel's more efficient machines, fueled by a heavier and cheaper fraction (you guessed it, diesel) and used above all for ships, trucks, and heavy machinery (for much more on this, see chapter 4 on globalization). Diffusion of these new prime movers was slow, and the US and Canada were the only two countries with high rates of car ownership prior to the Second World War.

Crude oil became a global fuel, and eventually the world's most important source of primary energy, thanks to the discoveries of giant oil fields in the Middle East and in the USSR—and, of course, also thanks to the introduction of large tankers. Some Middle Eastern giants were first drilled in the 1920s and 1930s (Iranian Gachsaran and Iraqi Kirkuk in 1927, Kuwaiti Burgan in 1937) but most of them were discovered after the war, including Ghawar (the world's largest) in 1948, Safaniya in 1951, and Manifa in 1957, all in Saudi Arabia. The largest Soviet discoveries were in 1948 (Romashkino in the Volga-Ural Basin) and in 1965 (Samotlor in Western Siberia).[42]

Crude oil's rise and relative retreat

Mass-scale car usage in Europe and Japan and the concurrent conversion of their economies from coal to crude oil, and later to natural gas, began only during the 1950s, as did the expansion of foreign trade and travel (including the first jetliners) and the use of petrochemical feedstocks for the synthesis of ammonia and plastics. Global oil extraction of crude oil doubled during the 1950s, and by 1964 crude oil surpassed coal as the world's most important fossil fuel, but although its output kept on rising, supply remained plentiful and so prices were falling. In constant (inflation-adjusted) monies, the world oil price was lower in 1950 than it was in 1940, lower in 1960 than in 1950—and lower still in 1970 than in 1960.[43]

Not surprisingly, demand was coming from all sectors. In real terms, crude oil was so cheap that there were no incentives to use it efficiently: American houses in regions with a cold climate, increasingly heated by oil furnaces, were built with single-glazed windows and without adequate wall insulation; the average efficiency of American cars actually declined between 1933 and 1973; and energy-intensive industries continued to operate by using inefficient processes.[44] Perhaps most notably, America's pace of replacing old open-hearth furnaces with superior oxygen furnaces to make steel was much slower than in Japan and Western Europe.

During the late 1960s, the already high American demand for oil rose by nearly 25 percent, and global demand increased by nearly 50 percent. European demand had nearly doubled between 1965 and 1973, and Japanese imports became about 2.3 times higher.[45] As mentioned, new discoveries of oil covered this surge in demand and oil was selling at what was essentially the same price as in 1950. This was too good to last, however. In 1950 the US still produced about 53 percent of the world's oil; by 1970, although still the largest producer, its share fell to less than 23 percent—and it was clear that the country would need increased imports—while the Organization of the Petroleum Exporting Countries (OPEC) produced 48 percent.

OPEC, set up in 1960 in Baghdad by just five countries in order to prevent further price reductions, had time on its side: it wasn't large enough to assert itself during the 1960s, but by 1970 its production share, combined with the retreat of American extraction (which peaked in 1970), made it impossible to ignore its demands.[46] In April 1972 the Texas Railroad Commission lifted its limits on the state's output and hence surrendered its control of the price that it had held since the 1930s. In 1971, Algeria and Libya began to nationalize their oil production, and Iraq followed in 1972, the same year that Kuwait, Qatar, and Saudi Arabia began their gradual takeover of their oilfields—which until that point had been in the hands of foreign corporations. Then in April 1973, the US ended its limits on the import of crude oil east of the Rocky Mountains. Suddenly, it was a sellers' market, and on October 1 1973 OPEC raised its posted price by 16 percent to $3.01/barrel, followed by an additional 17 percent rise by six Arab Gulf states and, after the Israeli victory over Egypt in Sinai in October 1973, it embargoed all oil exports to the US.

On January 1 1974, the Gulf states raised their posted price to $11.65/barrel, completing a 4.5-fold rise in the cost of this essential energy source in a single year—and this ended the era of rapid economic expansion that had been energized by cheap oil. From 1950 to 1973 the Western European economic product had nearly tripled, and the US GDP had more than doubled in that single generation. Between 1973 and 1975 the global economic growth rate dropped by about 90 percent, and as soon as the economies affected by higher oil prices began to adjust to these new realities—above all by impressive improvements in industrial energy efficiency—the fall of the Iranian monarchy and the takeover of Iran by a fundamentalist theocracy led to a second wave of oil price rises, from about $13 in 1978 to $34 in 1981, and to another 90 percent decline in the global rate of economic growth between 1979 and 1982.[47]

More than $30 a barrel was a demand-destroying price and by 1986 oil was again selling at just $13 a barrel, setting the stage for yet another round of globalization—this time centered on China, whose rapid modernization was driven by Deng Xiaoping's economic reforms and by massive foreign investment. Two generations later, only those who lived through those years of price and supply turmoil (or those, increasingly few, who studied their impact) appreciate how traumatic these two waves of price rises were. Consequences of the resulting economic reversals are still felt four decades later, because once demand for oil began to increase, many oil-saving measures remained in place and some—notably the transitions to more efficient industrial uses—kept on intensifying.[48]

In 1995, crude oil extraction finally surpassed the 1979 record and then continued to rise, meeting the demand of an economically reforming China as well as the rising demand elsewhere in Asia—but oil has not regained its pre-1975 relative dominance.[49] Its share of the global commercial primary energy supply fell from 45 percent in 1970 to 38 percent in the year 2000 and to 33 percent in 2019—and it is now certain that its further relative decline will continue as natural gas consumption and wind and solar electricity generation keep increasing. There are enormous opportunities to generate more electricity with photovoltaic cells and wind turbines, but there is a fundamental difference between systems that derive 20–40 percent of electricity from these intermittent sources (Germany and Spain are the best examples among large economies) and a national electricity supply that relies completely on these renewable flows.

In large, populous nations, the complete reliance on these renewables would require what we are still missing: either mass-scale, long-term (days to weeks) electricity storage that would back up intermittent electricity generation, or extensive grids of high-voltage lines to transmit electricity across time zones and from sunny and windy regions to major urban and industrial concentrations. Could these new renewables produce enough electricity to replace not only today's generation fueled by coal and natural gas, but also all the energy now supplied by liquid fuels to vehicles, ships, and planes by way of a complete electrification of transport? And could they really do so, as some plans now promise, in a matter of just two or three decades?

The many advantages of electricity

If energy, according to Feynman, is "that abstract thing," then electricity is one of its most abstract forms. You don't need a scientific understanding to have direct experience of several different types of energy, to distinguish their forms and take advantage of their conversions. Solid or liquid fuels (chemical energy) are tangible (a tree trunk, a lump of coal, a canister of gasoline), and their burning—be it in forest fires, in Paleolithic caves, in locomotives to produce steam, or in motor vehicles—releases heat (thermal energy). Falling and running waters are ubiquitous displays of gravitational and kinetic energy that are fairly easily converted to useful kinetic (mechanical) energy by building simple wooden waterwheels—and all it takes to convert wind's kinetic energy into mechanical energy for grinding grain or pressing oil seeds is a windmill and wooden gears to transfer the motion to millstones.

In contrast, electricity is intangible and we can't get an intuitive sense of it in the same way as we do with fuels. But its effects can be seen in static electricity, sparks, lightning; small currents can be felt, and currents above 100 milliamperes may be deadly. Common definitions of electricity are not instinctively accessible, they require a prior knowledge of other functional terms such as "electrons," "flow," "charge," and "current." Although Feynman, in the opening volume of his magisterial *Lectures on Physics*, was quite perfunctory—"there is electrical energy, which has to do with pushing and pulling by electric charges"—when he returned to the topic in detail in the second volume, dealing with mechanical and electrical energies and with steady currents, he did so by deploying calculus.[50]

For most of its inhabitants, the modern world is full of black boxes, devices whose internal workings remain—to different degrees—a mystery to their users. Electricity can be thought of as a ubiquitous and ultimate black box system: although many people have a fairly good understanding of what goes in (combustion of fossil fuel in a large thermal plant; falling water in a hydro station;

solar radiation absorbed by a photovoltaic cell; the splitting of uranium in a reactor) and everybody benefits from what comes out (light, heat, motion), only a minority fully understand what goes on inside the generating plants, transformers, transmission lines, and final-use devices.

Lightning, electricity's most common natural demonstration, is too powerful, too short-lived (only a fraction of a second), and too destructive to be (ever?) tapped for productive use. And while anybody can produce minute quantities of static electricity by rubbing suitable materials or use small batteries that can last, without recharging, for hours of light-duty service in flashlights and portable electronics, generating electricity for mass-scale commercial use is a costly and complicated undertaking. Its distribution from where it is generated to the places and regions of its largest use—to cities, industries, and electrified forms of rapid transportation—is equally complicated: it requires transformers and extensive grids of high-voltage transmission lines and, after further transformation, distribution by low-voltage overhead or underground wires to billions of consumers.

And even in this era of high-tech electronic miracles, it is still impossible to store electricity affordably in quantities sufficient to meet the demand of a medium-sized city (500,000 people) for only a week or two, or to supply a megacity (more than 10 million people) for just half a day.[51] But despite these complications, high costs, and technical challenges, we have been striving to electrify modern economies, and this quest for ever-higher electrification will continue because this form of energy combines many unequaled advantages. Most obviously, at the point of its final consumption, electricity's use is always effortless and clean, and the majority of the time it is also exceptionally efficient. With just the flip of a switch, push of a button, or adjustment of a thermostat (now often requiring only a hand signal or voice command), electric lights and motors or electric heaters and coolers are turned on—with no bulky fuel storages, no laborious carrying and stoking, no dangers of incomplete combustion (emitting poisonous carbon monoxide), and no cleaning of lamps or stoves or furnaces.

Electricity is the best form of energy for lighting: it has no competitor on any scale of private or public illumination, and very few innovations have produced such an impact on modern civilization as has the ability to remove the limits of daylight and to illuminate the night.[52] All previous alternatives, from ancient wax candles and oil lamps to early industrial gas lights and kerosene cylinders, were feeble, costly, and highly inefficient. The most telling comparison of light sources is in terms of their luminous efficacy—their ability to produce a visual signal, measured as the quotient of the total luminous flux (the total amount of energy put out by a source, in lumens) and the source's power (in watts). When setting the luminous efficacy of candles as equal to 1, coal gas lights in the early industrial cities produced 5–10 times more; before the First World War electric light bulbs with tungsten filaments emitted up to 60 times more; today's best fluorescent lights produce about 500 times as much; and sodium lamps (used for outdoor lighting) are up to 1,000 times more efficacious.[53]

It is impossible to decide which class of electricity converters has had a greater impact—lights or motors. The conversion of electricity into kinetic energy by electric motors first revolutionized nearly every sector of industrial production and later penetrated every household niche. Less demanding manual tasks and those that deployed steam engines to lift, press, cut, weave, and other industrial operations were almost completely electrified. In the US this took place within just four decades after the introduction of the first AC electric motors.[54] By 1930, electric drive had nearly doubled American manufacturing productivity, and had done so again by the late 1960s.[55] Concurrently, electric motors began their gradual conquest of rail transportation, beginning with electric streetcars and then with passenger trains.

The service sector now dominates all modern economies, and its operation is completely dependent on electricity. Electric motors power elevators and escalators, air-condition buildings, open doors, and compact garbage. They are also indispensable for e-commerce, as they power mazes of conveyor belts in giant warehouses. But the most ubiquitous units are never seen by people who rely on them every day. They are the tiny units activating mobile phone vibrators: the smallest ones measure less than 4 mm × 3 mm, their width being less than half the width of an average adult's pinky nail. You can see one only by dismantling your phone, or watching a video of that operation online.[56]

In some countries, virtually all rail transport is now electrified, and all high-speed trains (up to 300 km/h) are powered either by electric locomotives or by motors mounted in multiple locations, as is the case with Japan's pioneering Shinkansen launched in 1964.[57] And even basic car models now have between 20 and 40 small electric motors, with many more in expensive cars—adding to the vehicle's weight and increasing the drain on its batteries.[58] In households, besides lighting and powering all electronic devices—now commonly including security systems—electricity dominates mechanical tasks and supplies both heat and refrigeration in kitchens and energy for water heating, as well as heating for many houses.[59]

Without electricity, drinking water in all cities—as well as liquid and gaseous fossil fuels everywhere—would be unavailable. Powerful electric pumps feed water into the municipal supply, and they have an especially demanding task in cities with high commercial and residential densities where water must be lifted to a great height.[60] Electric motors run all the fuel pumps needed to move gasoline, kerosene, and diesel into tanks and wings. And while there may be plenty of natural gas in distribution gas pipelines—gas turbines are often used to move the fuel—in North America, where forced-air heating dominates, small electric motors operate fans that push the air heated by natural gas through the ducts.[61]

The long-term trend toward the electrification of societies (rising share of fuels converted to electricity rather than consumed directly) has been unmistakable. The new renewables—solar and wind, as opposed to hydroelectricity whose beginnings go back to 1882—will readily feed into this progression, but the history of electricity generation reminds us that many complications and complexities accompany the process; and that, despite its profound and rising importance, electricity still supplies only a relatively small share of final global energy consumption, just 18 percent.

Before you flip a switch

We need to go back to the industry's beginnings to appreciate its foundations, its infrastructure, and the legacy of these 140 years of development. Commercial electricity generation began in 1882, with three firsts. Two of them were the pioneering coal-fired generating stations designed by Thomas Edison (Holborn Viaduct in London began operating in January 1882; Pearl Street station in New York in September 1882), and the third was the first hydroelectric station (on the Fox River in Appleton, Wisconsin, also generating since September 1882).[62] Generation began to expand quickly during the 1890s, when alternating current (AC) transmission prevailed over the existing direct current networks, and when new designs of AC electric motors began to be adopted by industry and households. In 1900, less than 2 percent of the world's fossil fuel production was used to generate electricity; by 1950 that share was still less than 10 percent; it now stands at about 25 percent.[63]

The concurrent expansion of hydroelectric capacity accelerated during the 1930s, with large state-funded projects in the USA and the USSR, and reached new highs after the Second World War, culminating in the construction of record-size projects in Brazil (Itaipu, completed in 2007, 14 gigawatts) and China (Three Gorges, completed in 2012, 22.5 gigawatts).[64] Meanwhile, nuclear fission began to generate commercial electricity in 1956 at Britain's Calder Hall, saw its greatest expansion during the 1980s, peaked in 2006, and has since declined slightly to about 10 percent of global electricity generation.[65] Hydro generation accounted for nearly 16 percent in 2020; wind and solar added almost 7 percent; and the rest (about two-thirds) came from large central stations fueled mostly by coal and natural gas.

Not surprisingly, demand for electricity has been growing much faster than the demand for all other commercial energy: in the 50 years between 1970 and 2020, global electricity generation quintupled while the total primary energy demand only tripled.[66] And the growth of baseload generation—the minimum amount of electricity that has to be supplied on a daily, monthly, or annual basis—was further increased as progressively larger shares of populations moved to cities. Decades ago, American demand was lowest during summer nights, with shops and factories closed, public transport shut down, and all but a small share of the population asleep, with open windows. Now the windows are shut as air conditioners hum through the night to make sleep possible during hot, muggy weather; in large cities and megacities, many factories run two shifts and many shops and airports remain open 24 hours a day. Only COVID-19 stopped New York's subway operating 24/7, and the Tokyo subway sleeps for just five hours (the first train from Tokyo Station to Shinjuku leaves at 5:16 a.m., the last one at 0:20 a.m.).[67] Satellite nighttime images taken years apart show how street, parking, and building lights shine ever brighter over ever-larger areas that often join with nearby cities to form huge illuminated conurbations.[68]

A very high reliability of electricity supply—grid managers talk about the desirability of reaching six nines: with 99.9999 percent reliability there are only 32 seconds of interrupted supply in a year!—is imperative in societies where electricity powers everything from lights (be they in hospitals, along runways, or to indicate emergency escapes) to heart-lung machines and myriad industrial processes.[69] If the COVID-19 pandemic brought disruption, anguish, and unavoidable deaths, those effects would be minor compared to having just a few days of a severely reduced electricity supply in any densely populated region, and if prolonged for weeks nationwide it would be a catastrophic event with unprecedented consequences.[70]

Decarbonization: pace and scale

There is no shortage of fossil fuel resources in the Earth's crust, no danger of imminently running out of coal and hydrocarbons: at the 2020 level of production, coal reserves would last for about 120 years, oil and gas reserves for about 50 years, and continued exploration would transfer more of them from the resource to the reserve (technically and economically viable) category. Reliance on fossil fuels has created the modern world, but concerns about the relatively rapid rate of global warming have led to widespread calls for doing away with fossil carbon as expeditiously as possible. Ideally, the decarbonization of the global energy supply should proceed fast enough to limit average global warming to no more than 1.5°C (at worst 2°C). That, according to most climate models, would mean reducing net global CO_2 emissions to zero by 2050 and keeping them negative for the remainder of the century.

Notice the key qualifying adjective: the target is not total decarbonization but "net zero" or carbon neutrality. This definition allows for continued emissions to be compensated by (as yet non-existent!) large-scale removal of CO_2 from the atmosphere and its permanent storage underground, or by such temporary measures as the mass-scale planting of trees.[71] By 2020, setting net-zero goals for years ending in five or zero has become a me-too game: more than 100 nations have joined the lineup, ranging from Norway in 2030 and Finland in 2035 to the entire European Union, as well as Canada, Japan, and South Africa, in 2050, and China (the world's largest consumer of fossil fuels) in 2060.[72] Given the fact that annual CO_2 emissions from fossil fuel combustion surpassed 37 billion tons in 2019, the net-zero goal by 2050 will call for an energy transition unprecedented in both pace and scale. A closer look at its key components reveals the magnitude of the challenges.

Decarbonization of electricity generation can make the fastest progress, because installation costs per unit of solar or wind capacity can now compete with the least expensive fossil-fueled choices, and some countries have already transformed their generation to a considerable degree. Among large economies, Germany is the most notable example: since the year 2000, it has boosted its wind and solar capacity 10-fold and raised the share of renewables (wind, solar, and hydro) from 11 percent to 40 percent of total generation. Intermittency of wind and solar electricity poses no problems as long as these new renewables supply relatively small shares of the total demand, or as long as any shortfalls can be made up by imports.

As a result, many countries now produce up to 15 percent of all electricity from intermittent sources without any major adjustments, and Denmark shows how a relatively small and well-interconnected market can go far higher.[73] In 2019, 45 percent of its electricity came from wind generation, and this exceptionally high share can be sustained without any massive domestic reserve capacities, because any shortfalls can be readily made up by imports from Sweden (hydro and nuclear electricity) and Germany (electricity coming from many sources). Germany could not do the same: its demand is more than 20 times the Danish total, and the country must maintain a sufficient reserve capacity that could be activated when new renewables are dormant.[74] In 2019, Germany generated 577 terawatt-hours of electricity, less than 5 percent more than in 2000—but its installed generating capacity expanded by about 73 percent (from 121 to about 209 gigawatts). The reason for this discrepancy is obvious.

In 2020, two decades after the beginning of *Energiewende*, its deliberately accelerated energy transition, Germany still had to keep most of its fossil-fired capacity (89 percent of it, actually) in order to meet demand on cloudy and calm days. After all, in gloomy Germany, photovoltaic generation works on average only 11–12 percent of time, and the combustion of fossil fuels still produced nearly half (48 percent) of all electricity in 2020. Moreover, as its share of wind generation has increased, its construction of new high-voltage lines to transmit this electricity from the windy north to the southern regions of high demand has fallen behind. And in the US, where much larger transmission projects would be needed to move wind electricity from the Great Plains and solar electricity from the Southwest to high-demand coastal areas, hardly any long-standing plans to build these links have been realized.[75]

As challenging as such arrangements are, they rely on technically mature (and still improving) solutions—that is, on more efficient PV cells, large onshore and offshore wind turbines, and high-voltage (including long-distance direct current) transmission. If costs, permitting processes, and not-in-my-backyard sentiments were no obstacles, these techniques could be deployed fairly rapidly and economically. Moreover, the problems of intermittency of solar and wind generation could be resolved by renewed reliance on nuclear electricity generation. A nuclear renaissance would be particularly helpful if we cannot develop better ways of large-scale electricity storage soon.

We need very large (multi-gigawatt-hour) storage for big cities and megacities, but so far the only viable option to serve them is pumped hydro storage (PHS): it uses cheaper nighttime electricity to pump water from a low-lying reservoir to high-lying storage, and its discharge provides instantly available generation.[76] With renewably generated electricity, the pumping could be done whenever surplus solar or wind capacity is available, but obviously PHS can work only in places with suitable elevation differences and the operation consumes about a quarter of generated electricity for the uphill pumping of water. Other energy storages, such as batteries, compressed air, and supercapacitors, still have capacities orders of magnitude lower than needed by large cities, even for a single day's worth of storage.[77]

In contrast, modern nuclear reactors, if properly built and carefully run, offer safe, long-lasting, and highly reliable ways of electricity generation; as already noted, they are able to operate more than 90 percent of the time, and their lifespan can exceed 40 years. Still, the future of nuclear generation remains uncertain. Only China, India, and South Korea are committed to further expansion of their capacities. In the West, the combination of high capital costs, major construction delays, and the availability of less expensive choices (natural gas in the US, wind and solar in Europe) has made new fission capacities unattractive. Moreover, America's new small, modular, and inherently safe reactors (first proposed during the 1980s) have yet to be commercialized, and Germany, with its decision to abandon all nuclear generation by 2022, is only the most obvious example of Europe's widely shared, deep antinuclear sentiment (for the assessment of real nuclear generation risks, see chapter 5).

But this may not last: even the European Union now recognizes that it could not come close to its extraordinarily ambitious decarbonization target without nuclear reactors. Its 2050 net-zero emissions scenarios set aside the decades-long stagnation and neglect of the nuclear industry, and envisage up to 20 percent of all energy consumption coming from nuclear fission.[78] Notice that this refers to total primary energy consumption, not just to electricity. Electricity is only

18 percent of total final global energy consumption, and the decarbonization of more than 80 percent of final energy uses—by industries, households, commerce, and transportation—will be even more challenging than the decarbonization of electricity generation. Expanded electricity generation can be used for space heating and by many industrial processes now relying on fossil fuels, but the course of decarbonizing modern long-distance transportation remains unclear.

How soon will we fly intercontinentally on a wide-body jet powered by batteries? News headlines assure us that the future of flight is electric—touchingly ignoring the huge gap between the energy density of kerosene burned by turbofans and today's best lithium-ion (Li-ion) batteries that would be on board these hypothetically electric planes. Turbofan engines powering jetliners burn fuel whose energy density is 46 megajoules per kilogram (that's nearly 12,000 watt-hours per kilogram), converting chemical to thermal and kinetic energy—while today's best Li-ion batteries supply less than 300 Wh/kg, more than a 40-fold difference.[79] Admittedly, electric motors are roughly twice as efficient energy converters as gas turbines, and hence the effective density gap is "only" about 20-fold. But during the past 30 years the maximum energy density of batteries has roughly tripled, and even if we were to triple that again densities would still be well below 3,000 Wh/kg in 2050—falling far short of taking a wide-body plane from New York to Tokyo or from Paris to Singapore, something we have been doing daily for decades with kerosene-fueled Boeings and Airbuses.[80]

Moreover (as will be explained in chapter 3), we have no readily deployable commercial-scale alternatives for energizing the production of the four material pillars of modern civilization solely by electricity. This means that even with an abundant and reliable renewable electricity supply, we would have to develop new large-scale processes to produce steel, ammonia, cement, and plastics.

Not surprisingly, decarbonization outside of electricity generation has progressed slowly. Germany will soon generate half of its electricity from renewables, but during the two decades of *Energiewende* the share of fossil fuels in the country's primary energy supply has only declined from about 84 percent to 78 percent:

Germans like their unrestricted Autobahn speeds and their frequent intercontinental flying, and German industries hum on natural gas and oil.[81] If the country replicates its past record, then in 2040 its dependence on fossil fuels will still be close to 70 percent.

And what about countries that have not pushed renewables at extraordinary expense? Japan is the foremost example: in the year 2000 about 83 percent of its primary energy came from fossil fuels; in 2019 that share (due to the post-Fukushima loss of nuclear generation and the need for higher fuel imports) was 90 percent![82] And while the US has greatly reduced its dependence on coal—replaced by natural gas in electricity generation—the country's share of fossil fuels in primary energy supply was still 80 percent fossil in 2019. Meanwhile China's share of fossil fuels fell from 93 percent in the year 2000 to 85 percent in 2019—but this relative decline was accompanied by a near tripling of the country's fossil fuel demand. The economic rise of China was the main reason why the global consumption of fossil fuels rose by about 45 percent during the first two decades of the 21st century, and why, despite extensive and expensive expansion of renewable energies, the share of fossil fuels in the world's primary energy supply fell only marginally, from 87 percent to about 84 percent.[83]

Annual global demand for fossil carbon is now just above 10 billion tons a year—a mass nearly five times more than the recent annual harvest of all staple grains feeding humanity, and more than twice the total mass of water drunk annually by the world's nearly 8 billion inhabitants—and it should be obvious that displacing and replacing such a mass is not something best handled by government targets for years ending in zero or five. Both the high relative share and the scale of our dependence on fossil carbon make any rapid substitutions impossible: this is not a biased personal impression stemming from a poor understanding of the global energy system – but a realistic conclusion based on engineering and economic realities.

In contrast to recent hasty political pledges, these realities have been recognized by all carefully considered long-term energy supply scenarios. The Stated Policies Scenario published by the International Energy Agency (IEA) in 2020 sees the share of fossil fuels declining from 80 percent of the total global demand in 2019 to 72 percent by 2040, while the IEA's Sustainable Development Scenario (its most aggressive decarbonization scenario so far, allowing for substantially accelerated global decarbonization) envisages fossil fuels supplying 56 percent of the global primary energy demand by 2040, making it highly improbable that this high share could be cut close to zero in a single decade.[84]

Certainly, the affluent world—given its wealth, technical capabilities, high level of per capita consumption and the concomitant level of waste—can take some impressive and relatively rapid decarbonization steps (to put it bluntly, it should do with using less energy of any kind). But that is not the case with the more than 5 billion people whose energy consumption is a fraction of those affluent levels, who need much more ammonia to raise their crop yields to feed their increasing populations, and much more steel and cement and plastics to build their essential infrastructures. What we need is to pursue a steady reduction of our dependence on the energies that made the modern world. We still do not know most of the particulars of this coming transition, but one thing remains certain: it will not be (it cannot be) a sudden abandonment of fossil carbon, nor even its rapid demise—but rather its gradual decline.[85]

2. Understanding Food Production

Eating Fossil Fuels

Securing a sufficient quantity and nutritional variety of food is the existential imperative for every species. During their long evolution, our hominin ancestors evolved key physical advantages—erect posture, bipedalism, and relatively large brains—that set them apart from their simian ancestors. This combination of traits enabled them to become better scavengers, collectors of plants, and hunters of small animals.

Early hominins had only the simplest stone tools (hammerstones, choppers), which were useful for butchering animals, but they had no artifacts to aid hunting and catching. They could easily kill injured or sick animals and small, slower-moving mammals, but most of the meat of larger prey came from scavenging kills made by wild predators.[1] The eventual deployment of long spears, shafted axes, bows and arrows, woven nets, baskets, and fishing rods made it possible to hunt and catch a wide variety of species. Some groups—most notably the mammoth hunters of the Upper Paleolithic (this age ended about 12,000 years ago)—mastered the slaughter of large beasts, while many coastal dwellers became accomplished fishers: some even used boats to kill small migrating whales.

The transition from foraging (hunting and collecting) to sedentary living, supported by early agriculture and the domestication of several mammalian and avian species, resulted in a generally more predictable, but still often unreliable, food supply that was able to support much higher population densities than was the case for earlier groups—but this didn't necessarily mean better average nutrition. Foraging in arid environments could require an area of more than 100 square kilometers to support a single family. For today's Londoners, that is roughly the distance from Buckingham Palace to the Isle of Dogs; for New Yorkers, that's how a seagull flies from Manhattan's tip to the middle of Central Park: a lot of ground to cover simply to survive.

In more productive regions, population densities could rise to as many as 2–3 people per 100 hectares (equal to about 140 standard soccer fields).[2] The only foraging societies with high population densities were coastal groups (most notably in the Pacific Northwest), who had access to annual fish migrations and plentiful opportunities to hunt aquatic mammals: reliable supply of high-protein, high-fat food allowed some of them to switch to sedentary lives in large communal wooden homes, and left them with spare time to carve impressive totem poles. In contrast, early agriculture, where the just-domesticated crops were harvested, meant that more than one person per hectare of cultivated land could be fed.

Unlike the foragers who might have gathered scores of wild species, practitioners of early agriculture had to narrow the variety of the plants they cultivated, as a few staple crops (wheat, barley, rice, corn, legumes, potatoes) dominated typical, overwhelmingly plant-based, diets—but these crops could support population densities that were two or three orders of magnitude higher than in foraging societies. In ancient Egypt, the density rate rose from about 1.3 people per hectare of cultivated land during the predynastic period (pre–3150 BCE) to about 2.5 people per hectare 3,500 years later, when the country was a province of the Roman Empire.[3] This is equivalent to needing an area of 4,000 square meters to feed one person—or almost exactly six tennis courts. But this high production density was (due to the Nile's reliable annual flooding) an exceptionally good performance.

Over time, and very slowly, preindustrial rates of food production rose even higher—but rates of 3 people per hectare were not achieved until the 16th century, and only then in intensively cultivated regions of Ming China; in Europe they remained below 2 people per hectare until the 18th century. This stagnation, or at least very slow gains, in feeding capacity during the long course of preindustrial history meant that until a few generations ago only a small share of well-fed elites did not have to worry about having enough to eat. Even during the occasional years of above-average harvests, typical diets remained monotonous, and malnutrition and undernutrition were common.

Harvests could fail, and crops were often destroyed in wars—famine was a regular occurrence. As a result, no recent transformation—such as increased personal mobility or a greater range of private possessions—has been so existentially fundamental as our ability to produce, year after year, a surfeit of food. Now most people in affluent and middle-income countries worry about what (and how much) is best to eat in order to maintain or improve their health and extend their longevity, not whether they will have enough to survive.

There are still significant numbers of children, adolescents, and adults who experience food shortages, particularly in the countries of sub-Saharan Africa, but during the past three generations their total has declined from the world's majority to less than 1 in 10 of the world's inhabitants. The United Nations' Food and Agricultural Organization (FAO) estimates that the worldwide share of undernourished people decreased from about 65 percent in 1950 to 25 percent by 1970, and to about 15 percent by the year 2000. Continued improvements (with fluctuations caused by temporary national or regional setbacks due to natural disasters or armed conflicts) lowered the rate to 8.9 percent by 2019—which means that rising food production reduced the malnutrition rate from 2 in 3 people in 1950 to 1 in 11 by 2019.[4]

This impressive achievement is even more noteworthy if expressed in a way that accounts for the intervening large-scale increase of the global population, from about 2.5 billion people in 1950 to 7.7 billion in 2019. The steep reduction in global undernutrition means that in 1950 the world was able to supply adequate food to about 890 million people, but by 2019 that had risen to just over 7 billion: a nearly eightfold increase in absolute terms!

What explains this impressive achievement? Answering that it must be due to higher crop yields is a truism. Saying that the increase has been the combined effect of better crop varieties, agricultural mechanization, fertilization, irrigation, and crop protection correctly describes the changes in key inputs—but it still misses the fundamental explanation. Modern food production, be it field cultivation of crops or the capture of wild marine species, is a peculiar hybrid dependent on two different kinds of energy. The first, and most obvious, is the Sun. But we also need the now indispensable input of fossil fuels, and the electricity produced and generated by humans.

When asked to give common examples of our reliance on fossil fuels, inhabitants of the colder parts of Europe and North America will think immediately about the natural gas used to heat their houses. People everywhere will point out the combustion of liquid fuels that power most of our transportation but the modern world's most important—and fundamentally existential—dependence on fossil fuels is their direct and indirect use in the production of our food. Direct use includes fuels to power all field machinery (mostly tractors, combines, and other harvesters), the transportation of harvests from fields to storage and processing sites, and irrigation pumps. Indirect use is much broader, taking into account the fuels and electricity used to produce agricultural machinery, fertilizers, and agrochemicals (herbicides, insecticides, fungicides), and other inputs ranging from glass and plastic sheets for greenhouses, to global positioning devices that enable precision farming.

The fundamental energy conversion producing our food has not changed: as always, we are eating, whether directly as plant foods or indirectly as animal foodstuffs, products of photosynthesis—the biosphere's most important energy conversion, powered by solar radiation. What has changed is the intensity of our crop, and animal, production: we could not harvest such abundance, and in such a highly predictable manner, without the still-rising inputs of fossil fuels and electricity. Without these anthropogenic energy subsidies, we could not have supplied 90 percent of humanity with adequate nutrition and we could not have reduced global malnutrition to such a degree, while simultaneously steadily decreasing the amount of time and the area of cropland needed to feed one person.

Agriculture—growing food crops for people and feed for animals—must be energized by solar radiation, specifically by the blue and red parts of the visible spectrum.[5] Chlorophylls and carotenoids, light-sensitive molecules in plant cells, absorb light at these wavelengths and use it to power photosynthesis, a multi-step sequence of chemical reactions that combines atmospheric carbon dioxide and water—as well as small amounts of elements including, notably, nitrogen and phosphorus—to produce new plant mass for grain, legume, tuber, oil, and sugar crops. Part of these harvests is fed to domestic animals to produce meat, milk, and eggs, and additional animal foods come from mammals that graze on grasses and aquatic species whose growth depends ultimately on phytoplankton, the dominant plant mass produced by aquatic photosynthesis.[6]

This has always been so, from the very beginnings of settled cultivation going back some 10 millennia—but two centuries ago the addition of non-solar forms of energy began to affect the crop production and later also the capture of wild marine species. Initially this impact was marginal, and it became notable only in the early decades of the 20th century.

To trace the evolution of this epochal shift, we'll look next at the past two centuries of American wheat production. However, I could quite easily have chosen English or French wheat yields, or Chinese or Japanese rice yields; while agricultural advances may have happened at different times in cultivated parts of North America, Western Europe, and East Asia, there is nothing unique about this comparative sequence that is based on US data.

Three valleys, two centuries apart

We'll start in the Genesee Valley, western New York, in 1801. The new republic is in the 26th year of its existence and yet American farmers grow bread wheat not just the same way their ancestors did before they emigrated from England to British North America a few generations ago, but in a manner not too different from practices in ancient Egypt more than two millennia ago.

The sequence begins with two oxen harnessed to a wooden plow whose cutting edge is shod with an iron plate. Seed, saved from the previous year's crop, is sown by hand, and brush harrows are used to cover it up. Putting the crop in takes about 27 hours of human labor for every seeded hectare.[7] And the most laborious tasks are still to come. The crop is harvested by cutting with sickles; cut stalks are bundled and tied manually in sheaves, and they are stacked upright (to make shocks or stooks) and left to dry. The sheaves are then hauled to a barn and threshed by flailing them on a hard floor, straw is stacked, and grain is winnowed (separated from the chaff), measured, and put into sacks. Securing the crop takes at least 120 hours of human labor per hectare.

The complete production sequence demands about 150 hours of human labor per hectare, as well as about 70 ox-hours. The yield is just one ton of grain per hectare, and of that at least 10 percent has to be set aside as seed for the next year's crop. Altogether, it takes about 10 minutes of human labor to produce a kilogram of wheat, and that would, with wholegrain flour, yield 1.6 kilograms (two loaves) of bread. This is laborious, slow, and low-yielding farming—but it is completely solar, and no other energy inputs are required beyond the Sun's radiation: the crops produce food for people and feed for animals; trees yield wood for cooking and heating; and wood is also used to make metallurgical charcoal for smelting iron ores and producing small metal objects including plow plates, sickles, scythes, knives, and strakes to cover wooden wagon wheels. In modern parlance, we would say that this farming requires no non-renewable (fossil fuel) energy inputs and only a minimum of non-renewable material subsidies (iron components, stones for gristmills), and that the production of both crops and materials relies solely on renewable energies deployed through the exertion of human and animal muscles.

A century later, in 1901, most of the country's wheat comes from the Great Plains and so we move to the Red River Valley, in eastern North Dakota. The Great Plains have been settled and industrialization has made enormous advances during the past two generations—although wheat farming still relies on draft animals, the wheat growing on large Dakota farms is highly mechanized. Teams of four powerful horses pull gang (multi-share) steel plows and harrows, mechanical seed drills are used for planting, mechanical harvesters cut the stalks and bind the sheaves, and only the stooking is done manually. Sheaves are hauled to stacks and fed to threshing machines powered by steam engines, and grain is taken to granaries. The entire sequence takes less than 22 hours per hectare, about 1/7 of the time it did in 1801.[8] In this extensive cultivation, large areas make up for low yields: yields remain low at 1 ton per hectare but the investment of human labor is only about 1.5 minutes per kilogram of grain (compared to 10 minutes in 1801), while the use of draft animals adds up to about 37 horse-hours per hectare, or more than 2 minutes per kilogram of grain.

This is a new, hybrid kind of farming, as the indispensable solar input is augmented by non-renewable anthropogenic energies derived overwhelmingly from coal. The new arrangement requires more animal labor than human labor, and as working horses (and mules in the American South) need grain feed—mainly oats—as well as fresh grass and hay, their large numbers make substantial demands on the country's crop production: about one-quarter of all American farmland is devoted to growing fodder for draft animals.[9]

High-productivity harvests are possible thanks to increasing infusions of fossil energies. Coal is used to make metallurgical coke charged into blast furnaces, and cast iron is converted to steel in open hearth furnaces (see chapter 3). Steel is needed for agricultural machinery as well as for making steam engines, rails, wagons, locomotives, and ships. Coal also powers steam engines and produces the heat and electricity required to manufacture plows, drills, harvesters (also the first combines), wagons, and silos, and to operate railroads and ships that distribute the grain to its final consumers. Inorganic fertilizers are making their first inroads with the imports of Chilean nitrates and with the application of phosphates mined in Florida.

In 2021, Kansas is the country's leading wheat-growing state and so we move to the Arkansas River Valley. In this heart of American wheat country, farms are now commonly three to four times larger than they were a century ago[10]—and yet most of the field work is done by only one or two people operating large machinery. The US Department of Agriculture stopped counting draft animals in 1961, and field work is now dominated by powerful tractors—many models have more than 400 horsepower and eight giant tires—pulling wide implements such as steel plows (with a dozen or more shares), seeders, and fertilizer applicators.[11]

Seed comes from certified growers, and young plants receive optimum amounts of inorganic fertilizers—above all, plenty of nitrogen applied as ammonia or urea—and targeted protection against insects, fungi, and competing weeds. Harvesting, and the concurrent threshing, is done by large combines that transfer grain directly to trucks to be transported to storage silos and sold around the country, or shipped to Asia or Africa. Producing wheat now takes less than two hours of human labor per hectare (compared to 150 hours in 1801), and with yields of around 3.5 tons per hectare this translates to less than two seconds per kilogram of grain.[12]

Many people nowadays admiringly quote the performance gains of modern computing ("so much data") or telecommunication ("so much cheaper")—but what about harvests? In two centuries, the human labor to produce a kilogram of American wheat was reduced from 10 minutes to less than two seconds. This is how our modern world really works. And as mentioned, I could have done similarly stunning reconstructions of falling labor inputs, rising yields, and soaring productivity for Chinese or Indian rice. The time frames would be different but the relative gains would be similar.

Most of the admired and undoubtedly remarkable technical advances that have transformed industries, transportation, communication, and everyday living would have been impossible if more than 80 percent of all people had to remain in the countryside in order to produce their daily bread (the share of the US population who were farmers in 1800 was 83 percent) or their daily bowl of rice (in Japan, close to 90 percent of people lived in villages in 1800). The road to the modern world began with inexpensive steel plows and inorganic fertilizers, and a closer look is needed to explain these indispensable inputs that have made us take a well-fed civilization for granted.

What goes in

Preindustrial farming done with human and animal labor and with simple wooden and iron tools had the Sun as the only source of energy. Today, as ever, no harvests would be possible without Sun-driven photosynthesis, but the high yields produced with minimal labor inputs and hence with unprecedented low costs would be impossible without direct and indirect infusions of fossil energies.

Some of these anthropogenic energy inputs are coming from electricity, which can be generated from coal or natural gas or renewables, but most of them are liquid and gaseous hydrocarbons supplied as machine fuels and raw materials.

Machines consume fossil energies directly as diesel or gasoline for field operations including the pumping of irrigation water from wells, for crop processing and drying, for transporting the harvests within the country by trucks, trains, and barges, and for overseas exports in the holds of large bulk carriers. Indirect energy use in making those machines is far more complex, as fossil fuels and electricity go into making not only the steel, rubber, plastics, glass, and electronics but also assembling these inputs to make tractors, implements, combines, trucks, grain dryers, and silos.[13]

But the energy required to make and to power farm machinery is dwarfed by the energy requirements of producing agrochemicals. Modern farming requires fungicides and insecticides to minimize crop losses, and herbicides to prevent weeds from competing for the available plant nutrients and water. All of these are highly energy-intensive products but they are applied in relatively small quantities (just fractions of a kilogram per hectare).[14] In contrast, fertilizers that supply the three essential plant macronutrients—nitrogen, phosphorus, and potassium—require less energy per unit of the final product but are needed in large quantities to ensure high crop yields.[15]

Potassium is the least costly to produce, as all it takes is potash (KCl) from surface or underground mines. Phosphatic fertilizers begin with the excavation of phosphates, followed by their processing to yield synthetic superphosphate compounds. Ammonia is the starting compound for making all synthetic nitrogenous fertilizers. Every crop of high-yielding wheat and rice, as well as of many vegetables, requires more than 100 (sometimes as much as 200) kilograms of nitrogen per hectare, and these high needs make the synthesis of nitrogenous fertilizers the most important indirect energy input in modern farming.[16]

Nitrogen is needed in such great quantities because it is in every living cell: it is in chlorophyll, whose excitation powers photosynthesis; in the nucleic acids DNA and RNA, which store and process all genetic information; and in amino acids, which make up all the proteins required for the growth and maintenance of our tissues. The element is abundant—it makes up nearly 80 percent of the atmosphere, organisms live submerged in it—and yet it is a key limiting factor in crop productivity as well as in human growth. This is one of the great paradoxical realities of the biosphere and its explanation is simple: nitrogen exists in the atmosphere as a non-reactive molecule (N_2), and only a few natural processes can split the bond between the two nitrogen atoms and make the element available to form reactive compounds.[17]

Lightning will do it: it produces nitrogen oxides, which dissolve in rain and form nitrates, and then forests, fields, and grasslands get fertilizer from above—but obviously this natural input is too small to produce crop harvests to feed the world's nearly 8 billion people. What lightning can do with tremendous temperatures and pressures, an enzyme (nitrogenase) can do in normal conditions: it is produced by bacteria associated with the roots of leguminous plants (pulses, as well as some trees) or that live freely in soil or in plants. Bacteria attached to the roots of leguminous plants are responsible for most natural nitrogen fixation—that is, for the cleavage of non-reactive N_2 and for the incorporation of nitrogen into ammonia (NH_3), a highly reactive compound that is readily converted into soluble nitrates and can supply plants with their nitrogen needs in return for organic acids synthesized by the plants.

As a result, leguminous food crops, including soybeans, beans, peas, lentils, and peanuts, are able to provide (fix) their own nitrogen supply, as can such leguminous cover crops as alfalfa, clovers, and vetches. But no staple grains, no oil crops (except for soybeans and peanuts), and no tubers can do that. The only way for them to benefit from the nitrogen-fixing abilities of legumes is to rotate them with alfalfa, clovers, or vetches, grow these nitrogen fixers for a few months, and then plow them under so the soils are replenished with reactive nitrogen to be picked up by the succeeding wheat, rice, or potatoes.[18] In traditional agricultures, the only other option to enrich soil nitrogen stores was to collect and apply human and animal wastes. But this is an inherently laborious and inefficient way to supply the nutrient. These wastes have very low nitrogen content and they are subject to volatilization losses (the conversion of liquids to gases—the ammonia smell from manure can be overpowering).

In preindustrial cropping, the wastes had to be collected in villages, towns, and cities, fermented in heaps or pits and—because of their low nitrogen content—applied to fields in massive amounts, commonly 10 tons per hectare but sometimes up to 30 tons (the latter mass being equivalent to 25–30 small European cars), in order to provide the needed nitrogen. Not surprisingly, this was commonly the most time-consuming task in traditional farming, claiming at least a fifth, and as much as a third, of all (human and animal) labor in cropping. Recycling organic wastes is hardly a topic addressed by famous novelists, but Émile Zola, always a complete realist, captured its importance when he described Claude, a young Parisian painter who "had quite a liking for manure." Claude volunteers to toss into the pit "the scourings of markets, the refuse that fell from that colossal table, remained full of life, and returned to the spot where the vegetables had previously sprouted . . . They rose again in fertile crops, and once more went to spread themselves out upon the market square. Paris rotted everything, and returned everything to the soil, which never wearied of repairing the ravages of death."[19]

But at what cost of human toil! This great nitrogen barrier to higher crop yields was nudged only during the 19th century with the mining and export of Chilean nitrates, the first inorganic nitrogenous fertilizer. The barrier was then broken decisively with the invention of ammonia synthesis by Fritz Haber in 1909 and with its rapid commercialization (ammonia was first shipped in 1913), but subsequent production grew slowly and the widespread application of nitrogenous fertilizers had to wait until after the Second World War.[20] New high-yielding varieties of wheat and rice introduced during the 1960s could not express their full yield potential without synthetic nitrogenous fertilizers. And the great productivity shift known as the Green Revolution could not have taken place without this combination of better crops and higher nitrogen applications.[21]

Since the 1970s, the synthesis of nitrogenous fertilizers has undoubtedly been the *primus inter pares* among agricultural energy subsidies—but the full scale of this dependence is only revealed by looking at detailed accounts of the energy required to produce various common foodstuffs. I have chosen three of them to use as examples, and I picked them because of their nutritional dominance. Bread has been the staple of European civilization for millennia. Given the religious proscriptions on the consumption of pork and beef, chicken is the only universally favored meat. And no other vegetable (although botanically a fruit) surpasses the annual production of tomatoes, now grown not only as a field crop but increasingly in plastic or glass greenhouses.

Each of these foodstuffs has a different nutritional role (bread is eaten for its carbohydrates, chicken for its perfect protein, tomatoes for their vitamin C content) but none of them could be produced so abundantly, so reliably, and so affordably without considerable fossil fuel subsidies. Eventually, our food production will change, but for now, and for the foreseeable future, we cannot feed the world without relying on fossil fuels.

The energy costs of bread, chicken, and tomatoes

Given the enormous variety of breads, I'm going to stick to just a few varieties of leavened breads common in Western diets and now available in places ranging from West Africa (*outre-mer* domain of the French baguette) to Japan (every major department store has a French or German bakery). We have to start with wheat, and helpfully there is no shortage of studies that have attempted to quantify all fuel and electricity inputs and to compare them per harvested area or per unit of yield for different kinds of grain crops.[22] Grain cultivation is at the bottom of the energy subsidy ladder, needing relatively little compared to our other chosen foodstuffs, but as we shall see, it still needs a surprisingly large amount of energy.

Efficient American production of rain-fed wheat on the large fields of the Great Plains requires only about 4 megajoules per kilogram of grain. Because such a large share of this energy is in the form of diesel fuel refined from crude oil, the comparison might be more tangible in terms of equivalents rather than in standard energy units (joules).[23]

Moreover, expressing the needs for diesel fuel in terms of volumes per unit of edible product (be it 1 kilogram, a loaf of bread, or a meal) makes such energy subsidies more readily imaginable.

With diesel fuel containing 36.9 megajoules per liter, the typical energy cost of wheat from the Great Plains is almost exactly 100 milliliters (1 deciliter or 0.1 liters) of diesel fuel per kilogram—just a bit less than half of the US cup measurement.[24] I will use specific volume equivalents of diesel fuel to label individual foodstuffs with the energy embedded in their production.

Basic sourdough bread is the simplest kind of a leavened bread, the staple of European civilization: it contains just bread flour, water, and salt, and the leavening is made, of course, from flour and water. A kilogram of this bread will be about 580 grams of flour, 410 grams of water, and 10 grams of salt.[25] Milling—that is, removing the seed's bran, the outer layer—reduces the mass of milled grains by about 25 percent (a flour extraction rate of 72–76 percent).[26] This means that to get 580 grams of bread flour, we have to start with about 800 grams of whole wheat, whose production requires 80 milliliters of diesel fuel equivalent.

Milling the grain needs an equivalent of about 50 mL/kg to produce white bread flour, while published data for large-scale baking in modern efficient enterprises—consuming natural gas and electricity—indicate fuel equivalents of 100–200 mL/kg.[27] Growing the grain, milling it, and baking a 1-kilogram sourdough loaf thus requires an energy input equivalent of at least 250 milliliters of diesel fuel, a volume slightly larger than the American measuring cup. For a standard baguette (250 grams), the embedded energy equivalent is about 2 tablespoons of diesel fuel; for a large German *Bauernbrot* (2 kilograms), it would be about 2 cups of diesel fuel (less for a wholewheat loaf).

The real fossil energy cost is higher still, because only a small share of bread is now baked where it is bought. Even in France, neighborhood *boulangeries* have been disappearing and baguettes are distributed from large bakeries: energy savings from industrial-scale efficiency are negated by increased transportation costs, and the total cost (from growing and milling grain to baking in a large bakery and distributing bread to distant consumers) may have an equivalent energy consumption as high as 600 mL/kg!

But if the bread's typical (roughly 5:1) ratio of edible mass to the mass of embedded energy (1 kilogram of bread compared to about 210 grams of diesel fuel) seems uncomfortably high, recall that I have already noted that grains—even grains after processing and conversion into our favorite foods—are at the bottom of our food energy subsidy ladder. What would be the consequences of following such a dubious dietary recommendation, now pushed by some promoters under the misleading label of the "Paleolithic diet," as avoiding all cereals and switching instead to diets composed only of meat, fish, vegetables, and fruit?

Rather than tracing the energy cost of beef (a meat that has already been much maligned), I will instead quantify the energy burdens of the most efficiently produced meat—that of broilers reared in large barns in what have become known as CAFOs, central animal feeding operations. In the case of chicken, this means housing and feeding tens of thousands of birds in long rectangular structures where they are crowded in dimly lit spaces (the equivalent of a moonlit night) and fed for about seven weeks before being taken away for slaughter.[28] The US Department of Agriculture publishes statistics on the annual feeding efficiency of domestic animals, and over the past five decades these ratios (units of feed expressed in terms of corn grain per unit of live weight) show no downward trends for either beef or pork, but impressive gains for chicken.[29]

In 1950, 3 units of feed were needed per unit of live broiler weight; now that number is just 1.82, about a third of the rate for pigs and a seventh of the rate for cattle.[30] Obviously, the entire bird (including feathers and bones) is not eaten, and the adjustment for edible weight (about 60 percent of live weight) puts the lowest feed-to-meat ratio at 3:1. Producing one American chicken (whose average edible weight is now almost exactly 1 kilogram) needs 3 kilograms of grain corn.[31] Corn's efficient, rain-fed cultivation has high yields and relatively low energy costs—equivalent to about 50 milliliters of diesel fuel per kilogram of grain—but the energy cost of irrigated corn may be twice as high as that of rain-fed feed, and typical corn yields and feeding efficiencies around the world are lower than in the US. As a result, feed costs alone can be as low as 150 milliliters of diesel fuel per kilogram of edible meat, and as high as 750 mL/kg.

Further energy costs arise from a large-scale intercontinental trade in feedstuffs: it is dominated by the shipment of American corn and soybeans and the sale of Brazilian soybeans. Brazilian soybean cultivation requires the equivalent of 100 milliliters of diesel fuel per kilogram of grain, but trucking the crop from producing areas to ports and shipping it to Europe doubles the energy cost.[32] Growing broilers to slaughter weight also requires energy for heating, air conditioning, and maintaining the poultry houses, for supplying water and sawdust, and for removing and composting waste. These requirements vary widely with location (above all, due to summer air conditioning and winter heating), and hence when combined with the energy cost of delivered feed a wide range of volumes is produced—from 50 to 300 milliliters per kilogram of edible meat.[33]

The most conservative combined rate for feeding and rearing the birds would be thus an equivalent of about 200 milliliters of diesel fuel per kilogram of meat, but the values can go as high as 1 liter. Adding the energy needed for slaughtering and processing the birds (chicken meat is now overwhelmingly marketed as parts, not as whole broilers), retailing, storing and home refrigeration, and eventual cooking raises the total energy requirement for putting a kilogram of roasted chicken on dinner plates to at least 300–350 milliliters of crude oil: a volume equal to almost half a bottle of wine (and for the least efficient producers, to more than a liter).

The minima of 300–350 mL/kg is a remarkably efficient performance compared to the rates of 210–250 mL/kg for bread, and this is reflected in the comparably affordable prices of chicken: in US cities, the average price of a kilogram of white bread is only about 5 percent lower than the average price per kilogram of whole chicken (and wholewheat bread is 35 percent more expensive!), while in France a kilogram of standard whole chicken costs only about 25 percent more than the average price of bread.[34] This helps to explain the rapid rise of chicken to become the dominant meat in all Western countries (globally, pork still leads, thanks to China's enormous demand).

Given that vegans extol eating plants, and that the media have reported extensively on the high environmental cost of meat, you might think that gains in the energy cost of chicken have been surpassed by those in the cultivation and marketing of vegetables. You would be mistaken to think that. The opposite has been true, in fact, and there is no better example to illustrate these surprisingly high energy burdens than taking a close look at tomatoes. They have it all—an attractive color, a variety of shapes, smooth skin, and a juicy interior. Botanically, a tomato is the fruit of the *Lycopersicon esculentum*, a small plant native to Central and South America that was introduced to the rest of the world during the age of first European transatlantic sailings but which took generations to establish worldwide appeal.[35] Eaten out of hand, in soups, filled, baked, chopped, boiled, pureed into sauces, and added to countless salads and cooked dishes, it is now a global favorite embraced in countries ranging from its native Mexico and Peru to Spain, Italy, India, and China (now its largest producer).

Nutritional compendia praise its high vitamin C content: indeed, a large tomato (200 grams) can provide two-thirds of the daily recommended requirement for an adult.[36] But as with all fresh and juicy fruits, it is not eaten for its energy content; it is, overwhelmingly, just an appealingly shaped container of water, which comprises 95 percent of its mass. The remainder is mostly carbohydrate, a bit of protein, and a mere trace of fat.

Tomatoes can be grown anywhere with at least 90 days of warm weather, including the deck of a seaside cottage near Stockholm or in a garden on the Canadian Prairies (in both cases, from plants started indoors). Commercial cultivation is a different matter, however. As with all but a small share of the fruits and vegetables that are consumed in modern societies, tomato cultivation is a highly specialized affair and most of the varieties available in North American and European supermarkets come from only a few places. In the US it is California; in Europe it is Italy and Spain. In order to increase their yield, improve their quality, and reduce the intensity of energy inputs, tomatoes are increasingly grown in plastic-covered single- or multi-tunnel enclosures or in greenhouses—not only in Canada and the Netherlands but also in Mexico, China, Spain, and Italy.

This brings us back to fossil fuels and electricity. Plastics are a less expensive alternative to constructing multi-tunnel glass greenhouses, and the cultivation of tomatoes also requires plastic clips, wedges, and gutter arrangements. Where the plants are grown in the open, plastic sheets are used to cover the soil in order to reduce water evaporation and prevent weeds. The synthesis of plastic compounds relies on hydrocarbons (crude oil and natural gas), both for raw materials (feedstocks) and for the energy needed to produce them. Feedstocks include ethane and other natural gas liquids, and naphtha produced during the refining of crude oil. Natural gas is also used to fuel plastic production, and it is (as already noted) the most important feedstock—the source of hydrogen—for the synthesis of ammonia. Other hydrocarbons serve as feedstocks to produce protective compounds (insecticides and fungicides), because even plants inside glass or plastic greenhouses are not immune to pests and infections.

Expressing the annual operating costs of tomato cultivation in monies is done easily by adding up the expenditure on seedlings, fertilizers, agrochemicals, water, heating, and labor, and by prorating the costs of original structures and devices—metal supports, plastic covers, glass, pipes, troughs, heaters—that are in place for more than one year. But putting a comprehensive energy bill together is not that simple. Direct energy inputs are easy to quantify on the basis of electricity bills and gasoline or diesel fuel purchases, but calculating the indirect flows into the production of materials requires some specialized accounting, and usually some assumptions.

Detailed studies have quantified these inputs and multiplied them by their typical energy costs: for example, the synthesis, formulation, and packaging of 1 kilogram of nitrogenous fertilizer requires an equivalent of nearly 1.5 liters of diesel fuel. Not surprisingly, these studies show a wide range of totals, but one study—perhaps the most meticulous study of tomato cultivation in the heated and unheated multi-tunnel greenhouses of Almería in Spain—concluded that the cumulative energy demand of net production is more than 500 milliliters of diesel fuel (more than two cups) per kilogram for the former (heated) and only 150 mL/kg for the latter harvest.[37]

We get this high energy cost, in large part, because greenhouse tomatoes are among the world's most heavily fertilized crops: per unit area they receive up to 10 times as much nitrogen (and also phosphorus) as is used to produce grain corn, America's leading field crop.[38] Sulfur, magnesium, and other micronutrients are also used, as are chemicals protecting against insects and fungi. Heating is the most important direct use of energy in greenhouse cultivation: it extends the growing season and improves crop quality but, inevitably, when deployed in colder climates it becomes the single largest user of energy.

Plastic greenhouses located in the southernmost part of Almería province are the world's largest covered area of commercial cultivation of produce: about 40,000 hectares (think of a 20 km × 20 km square) and easily identifiable on satellite images—look for yourself on Google Earth. You can even take a ride on Google Street View, which offers an otherworldly experience of these low-elevation, plastic-covered structures. Under this sea of plastic, the Spanish growers and their local and immigrant African laborers produce annually (in temperatures often surpassing 40°C) nearly 3 million tons of early and out-of-season vegetables (tomatoes, peppers, green beans, zucchini, eggplant, melons) and some fruit, and export about 80 percent of it to EU countries.[39] A truck transporting a 13-ton load of tomatoes from Almería to Stockholm covers 3,745 kilometers and consumes about 1,120 liters of diesel fuel.[40] That works out to nearly 90 milliliters per kilogram of tomatoes, and transport, storage, and packing at the regional distribution centers as well as deliveries to stores raises that to nearly 130 mL/kg.

This means that when bought in a Scandinavian supermarket, tomatoes from Almería's heated plastic greenhouses have a stunningly high embedded production and transportation energy cost. Its total is equivalent to about 650 mL/kg, or more than five tablespoons (each containing 14.8 milliliters) of diesel fuel per medium-sized (125 gram) tomato! You can stage—easily and without any waste—a tabletop demonstration of this fossil fuel subsidy, by slicing a tomato of that size, spreading it out on a plate, and pouring over it 5–6 tablespoons of dark oil (sesame oil replicates the color well). When sufficiently impressed by the fossil fuel burden of this simple food, you can transfer the plate's contents to a bowl, add two or three additional tomatoes, some soy sauce, salt, pepper, and sesame seeds, and enjoy a tasty tomato salad. How many vegans enjoying the salad are aware of its substantial fossil fuel pedigree?

Diesel oil behind seafood

High agricultural productivities of modern societies have made hunting on land (the seasonal shooting of some wild mammals and birds) a marginal source of nutrition in all affluent societies. Wild meat, mostly illegally hunted, is still more common throughout sub-Saharan Africa, but with rapidly growing populations even there it has ceased to be a major source of animal protein. By contrast, marine hunting has never been practiced more widely and more intensively than it is today, as huge fleets of ships—ranging from large modern floating factories to decrepit small boats—scour the world's oceans for wild fish and crustaceans.[41]

As it turns out, capturing what the Italians so poetically call *frutti di mare* is the most energy-intensive process of food provision. Of course, not all seafood is difficult to catch, and harvesting many still-abundant species does not require long expeditions to the remote areas of the southern Pacific. Capturing such plentiful pelagic (living near the surface) species as anchovies and sardines or mackerel can be done with a relatively small energy investment—indirectly in constructing ships and making large nets, directly in the diesel fuel used for ship engines. The best accounts show energy expenditures as low as 100 mL/kg for their capture, an equivalent of less than half a cup of diesel fuel.[42]

If you want to eat wild fish with the lowest-possible fossil carbon footprint, stick to sardines. The mean for all seafood is stunningly high—700 mL/kg (nearly a full wine bottle of diesel fuel)—and the maxima for some wild shrimp and lobsters are, incredibly, more than 10 L/kg (and that includes a great deal of inedible shells!).[43] This means that just two skewers of medium-sized wild shrimp (total weight of 100 grams) may require 0.5–1 liters of diesel fuel to catch—the equivalent of 2–4 cups of fuel.

But, you will object, shrimp are now mostly aquacultured, and haven't these large-scale, industrial-type operations enjoyed the same advantages that we have exploited so successfully with broilers? Alas, no, because of their fundamental metabolic difference. Broilers are herbivores, and when in confinement their energy expenditure on activity is limited. Therefore feeding them suitable plant matter—now mostly a combination of corn- and soybean-based mixtures—will make them grow fast. Unfortunately, the marine species that people prefer to eat (salmon, sea bass, tuna) are carnivorous, and for their proper growth they need to be fed protein-rich fish meals and fish oil derived from catches of wild species such as anchovies, pilchards, capelin, herring, and mackerel.

Expanding aquaculture—whose total global output, freshwater and marine, is now closing in on the worldwide wild catch (in 2018 it was 82 million tons compared to 96 million tons of wild-caught species)—has eased the pressure on some overfished wild stocks of preferred carnivorous fishes, but it has intensified the exploitation of smaller herbivorous species whose growing harvests are needed to feed expanding aquaculture.[44] As a result, the energy costs of growing Mediterranean sea bass in cages (Greece and Turkey are its leading producers) are commonly equivalent to as much as 2–2.5 liters of diesel fuel per kilogram (a volume about the same as three bottles of wine)—that is, of the same order of magnitude as the energy costs of capturing similarly sized wild species.

As expected, only aquacultured herbivorous fish that grow well consuming plant-based feed—most notably, different species of Chinese carp (bighead, silver, black, and grass carp are the most common)—have a low energy cost, typically less than 300 mL/kg. But, traditional Christmas Eve dinners in Austria, Czech Republic, Germany, and Poland aside, carp is quite an unpopular culinary choice in Europe and it is barely eaten in North America, while demand for tuna, some species of which are now among the most endangered top marine carnivores, has been soaring thanks to the rapid worldwide adoption of sushi.

So, the evidence is inescapable: our food supply—be it staple grains, clucking birds, favorite vegetables, or seafood praised for its nutritious quality—has become increasingly dependent on fossil fuels. This fundamental reality is commonly ignored by those who do not try to understand how our world really works and who are now predicting rapid decarbonization. Those same people would be shocked to know that our present situation cannot be changed easily or rapidly: as we saw in the preceding chapter, the ubiquity and the scale of the dependence are too large for that.

Fuel and food

Several studies have traced the growth of food production's dependence on modern—overwhelmingly fossil—energy inputs, from their absence in the early 19th century to recent rates (ranging from less than 0.25 tons of crude oil per hectare in grain farming to 10 times as much in heated greenhouse cultivation).[45] Perhaps the best way to realize the rise and the extent of this global dependence is to compare the increase of external energy subsidies to the expansion of cultivated land and to the growth of the world's population. Between 1900 and the year 2000, the global population increased less than fourfold (3.7 times to be exact) while farmland grew by about 40 percent, but my calculations show that anthropogenic energy subsidies in agriculture increased 90-fold, led by energy embedded in agrochemicals and in fuels directly consumed by machinery.[46]

I have also calculated the relative global burden of this dependence. Anthropogenic energy inputs into modern field farming (including all transportation), fisheries, and aquaculture add up to only about 4 percent of recent annual global energy use. This may be a surprisingly small share, but it must be remembered that the Sun will always do most of the work of growing food, and that external energy subsidies target those components of the food system where the greatest returns can be expected by reducing or removing natural constraints—be it by fertilizing, irrigating, providing protection against insects, fungi and competing plants, or by promptly harvesting mature crops. The low share may also be seen as yet another convincing example of small inputs having disproportionately large consequences, not an uncommon finding in the behavior of complex systems: think of vitamins and minerals, needed daily in just milligrams (vitamin B6 or copper) or micrograms (vitamin D, vitamin B12) to keep bodies weighing tens of kilograms in good shape.

But the energy required for food production—field farming, animal husbandry, and seafood—is only a part of the total food-related fuel and electricity needs, and estimating the use in the entire food system results in much higher shares of the total supply. Our best data are available for the US, where, thanks to the prevalence of modern techniques and widespread economies of scale, the direct energy use in food production is now on the order of 1 percent of the total national supply.[47] But after adding the energy requirements of food processing and marketing, packaging, transportation, wholesale and retail services, household food storage and preparation, and away-from-home food and marketing services, the grand total in the US reached nearly 16 percent of the nation's energy supply in 2007 and now it is approaching 20 percent.[48] The factors driving these rising energy needs range from further consolidation of production—and hence growing transportation needs—and growing food import dependency, to more meals eaten away from home and more prepared (convenience) foods consumed at home.[49]

There are many reasons why we should not continue many of today's food-producing practices. Agriculture's major contribution to the generation of greenhouse gases is now the most-often cited justification for following a different path. But modern crop cultivation, animal husbandry, and aquaculture have many other undesirable environmental impacts, ranging from the loss of biodiversity to the creation of dead zones in coastal waters (for more on this see chapter 6)—and there are no good reasons for maintaining our excessive food production with its attendant food waste. So, many changes are clearly desirable, but how fast can they actually happen, and how radically can we reform our current ways in reality?

Can we go back?

Can we reverse at least some of these trends? Can the world of soon-to-be 8 billion people feed itself—while maintaining a variety of crop and animal products and the quality of prevailing diets—without synthetic fertilizers and without other agrochemicals? Could we return to purely organic cropping, relying on recycled organic wastes and natural pest controls, and could we do without engine-powered irrigation and without field machinery by bringing back draft animals? We could, but purely organic farming would require most of us to abandon cities, resettle villages, dismantle central animal feeding operations, and bring all animals back to farms to use them for labor and as sources of manure.

Every day we would have to feed and water our animals, regularly remove their manure, ferment it and then spread it on fields, and tend the herds and flocks on pasture. As seasonal labor demands rose and ebbed, men would guide the plows harnessed to teams of horses; women and children would plant and weed vegetable plots; and everybody would be pitching in during harvest and slaughter time, stooking sheaves of wheat, digging up potatoes, helping to turn freshly slaughtered pigs and geese into food. I do not foresee the organic green online commentariat embracing these options anytime soon. And even if they were willing to empty the cities and embrace organic earthiness, they could still produce only enough food to sustain less than half of today's global population.

The numbers to confirm all of the above are not difficult to marshal. The decline of human labor required to produce American wheat outlined earlier in this chapter is an excellent proxy for the overall impact that mechanization and agrochemicals have had on the size of the country's agricultural labor force. Between 1800 and 2020, we reduced the labor needed to produce a kilogram of grain by more than 98 percent—and we reduced the share of the country's population engaged in agriculture by the same large margin.[50] This provides a useful guide to the profound economic transformations that would have to take place with any retreat of agricultural mechanization and reduction in the use of synthetic agrochemicals.

The greater the reduction of these fossil fuel–based services, the greater the need for the labor force to leave the cities to produce food in the old ways. During the pre-1920 peak of US horse and mule numbers, one-quarter of the country's farmland was dedicated to growing feed for the more than 25 million American working horses and mules—and at that time US farms had to feed only about 105 million people. Obviously, feeding today's more than 330 million people by deploying "just" 25 million horses would be impossible. And without synthetic fertilizers, yields of food and feed crops dependent on the recycling of organic matter would be a fraction of today's harvest. Corn, America's largest crop, yielded less than 2 tons per hectare in 1920, and 11 tons per hectare in 2020.[51] Millions of additional draft animals would be needed to cultivate virtually all of the country's available farmland, and it would be impossible to find enough recyclable organic matter (and eager Claude-like manure-liking tossers!) or cultivate sufficiently large areas of green manures (rotating grain with alfalfa or clover) to match the nutrients supplied by today's applications of synthetic fertilizers.

This impossibility is best illustrated by a few sets of simple comparisons. Recycling of organic matter is always highly desirable, as it improves the structure of soil, increases its organic content, and provides energy for myriad soil microbes and invertebrates. But the very low nitrogen content of organic matter means that farmers have to apply very large quantities of straw or manure in order to supply enough of this essential plant nutrient to produce high crop yields. The nitrogen content of cereal straws (the most abundant crop residue) is always low, usually 0.3–0.6 percent; manure mixed with animal bedding (usually straw) contains only 0.4–0.6 percent; fermented human waste (China's so-called night soil) has just 1–3 percent; and manures applied to fields rarely contain more than 4 percent.

In contrast, urea, now the world's dominant solid nitrogenous fertilizer, contains 46 percent nitrogen, ammonium nitrate has 33 percent, and commonly used liquid solutions contain 28–32 percent, at least an order of magnitude more nitrogen-dense than recyclable wastes.[52] This means that to supply the same amount of the nutrient to growing crops, a farmer would have to apply anywhere between 10 and 40 times as much manure by mass—and in reality even more of it would be needed, as significant shares of nitrogenous compounds are lost due to volatilization, or dissolved in water and carried below the root level, with the aggregate losses of nitrogen from organic matter being almost always higher than those from a synthetic liquid or solid.

Moreover, there would be a more than commensurate claim on labor, as the handling, transporting, and spreading of manure is far more difficult than dealing with small, free-flowing granules that can be easily applied by mechanical spreaders or (as is done with urea in small Asian rice fields) simply by sowing it by hand. And regardless of the effort that might be put into organic recycling, the total mass of recyclable materials is simply too small to provide the nitrogen required by today's harvests.

Global inventory of reactive nitrogen shows that six major flows bring the element to the world's croplands: atmospheric deposition, irrigation water, plowing-under of crop residues, spreading of animal manures, nitrogen left in soil by leguminous crops, and application of synthetic fertilizers.[53]

Atmospheric deposition—mainly as rain and snow containing dissolved nitrates—and recycled crop residues (straws and plant stalks that are not removed from fields to feed animals or burned onsite) each contribute about 20 megatons of nitrogen per year. Animal manures applied to fields, mainly from cattle, pigs, and chickens, contain almost 30 megatons; a similar total is introduced by leguminous crops (green manure cover crops, as well as soybeans, beans, peas, and chickpeas); and irrigation water brings about 5 megatons—for a total of about 105 megatons of nitrogen per year. Synthetic fertilizers supply 110 megatons of nitrogen per year, or slightly more than half of the 210–220 megatons used in total. This means that at least half of recent global crop harvests have been produced thanks to the application of synthetic nitrogenous compounds, and without them it would be impossible to produce the prevailing diets for even half of today's nearly 8 billion people. While we could reduce our dependence on synthetic ammonia by eating less meat and wasting less food, replacing the global input of about 110 megatons of nitrogen in synthetic compounds by organic sources could be done only in theory.

Multiple constraints limit the recycling of manure produced by animals in confinement.[54] In traditional mixed farming, cattle, pig, and poultry manure from relatively small numbers of animals was directly recycled on adjacent fields. Producing meat and eggs in central animal feeding operations reduced this option: these enterprises generate such large quantities of waste that its application to fields would overload soils with nutrients within the radius where it would be profitable to spread it; presence of heavy metals and drug residues (from feed additives) is another problem.[55] Similar constraints apply to the expanded use of sewage sludge (biosolids) from modern human waste treatment plants. Waste's pathogens must be destroyed by fermentation and by high-heat sterilization, but such treatments do not kill all antibiotic-resistant bacteria and do not remove all heavy metals.

Grazing animals produce three times as much manure as do mammals and birds kept in confinement: the FAO estimates that they leave annually about 90 megatons of nitrogen in waste—but exploiting this large source is impractical.[56] Accessibility would limit any gathering of animal urine and excrement to a fraction of the hundreds of millions of hectares of pastures where these wastes are expelled by grazing cattle, sheep, and goats. Gathering it would be as prohibitively costly as its transportation to treatment points and then to crop fields. Moreover, intervening nitrogen losses would further reduce the already very low nitrogen content of such wastes before the nutrient could reach the fields.[57]

Another choice is to expand the cultivation of leguminous crops to produce 50–60 megatons of nitrogen per year, rather than about 30 megatons as they currently do—but only at a considerable opportunity cost. Planting more leguminous cover crops such as alfalfa and clover would boost nitrogen supply but would also reduce the ability to use one field to produce two crops in a year, a vital option for the still-expanding populations of low-income countries.[58] Growing more leguminous grains (beans, lentils, peas) would lower the overall food energy yields, because they yield far less than cereal crops and, obviously, this would reduce the number of people that could be supported by a unit of cultivated land.[59] Moreover, the nitrogen left behind by a soybean crop—commonly 40–50 kilograms of nitrogen per hectare—would be less than the typical American applications of nitrogenous fertilizers, which are now about 75 kg N/ha for wheat and 150 kg N/ha for grain corn.

Another obvious drawback of expanded rotations with leguminous crops is that in colder climates, where only a single crop can be grown in a year, cultivation of alfalfa or clover would preclude the annual planting of a food crop, while in warmer regions with double-cropping it would reduce the frequency of harvesting food crops.[60] While it might be possible in countries with small populations and plentiful farmland, it would, inevitably, reduce food-producing capacity in all places where double-cropping is common, including large parts of Europe and the North China Plain, the region that produces about half of China's grain.

Double-cropping is now practiced on more than a third of China's cultivated land, and more than a third of all rice comes from double-cropping in South China.[61] Consequently, the country would find it impossible to feed its now more than 1.4 billion people without this intensive cultivation that also requires record-level nitrogen applications. Even in traditional Chinese farming, famous for its high rate of organic recycling and for complex crop rotations, farmers in the most intensively cultivated regions could not supply more than 120–150 kg N/ha—and doing so required extraordinarily high labor inputs, with (as already stressed) manure collection and application being the most time-consuming.

Even so, such farms could produce only overwhelmingly vegetarian diets for 10–11 people per hectare. In contrast, China's most productive double-cropping depends on applications of synthetic nitrogenous fertilizers averaging more than 400 kg N/ha, and it can produce enough to feed 20–22 people whose diets contain about 40 percent animal and 60 percent plant protein.[62] Global crop cultivation supported solely by the laborious recycling of organic wastes and by more common rotations is conceivable for a global population of 3 billion people consuming largely plant-based diets, but not for nearly 8 billion people on mixed diets: recall that synthetic fertilizers now supply more than twice as much nitrogen as all recycled crop residues and manures (and given the higher losses from organic applications, the effective multiple is actually closer to three!).

Doing with less—and doing without

But none of this means that major shifts in our dependence on fossil fuel subsidies in food production are impossible. Most obviously, we could reduce our crop and animal production—and the attendant energy subsidies—if we wasted less food. In many low-income countries, poor crop storage (making grains and tubers vulnerable to rodents, insects, and fungi) and the absence of refrigeration (accelerating the spoilage of dairy products, fish, and meat) wastes too much food even before it reaches its markets. And in affluent countries, food chains are longer and opportunities for inadvertent food losses arise at every step.

Even so, the well-documented global food losses have been excessively high, mostly because of an indefensible difference between output and actual needs: daily average per capita requirements of adults in largely sedentary affluent populations are no more than 2,000–2,100 kilocalories, far below the actual supplies of 3,200–4,000 kilocalories.[63] According to the FAO, the world loses almost half of all root crops, fruits, and vegetables, about a third of all fish, 30 percent of cereals, and a fifth of all oilseeds, meat, and dairy products—or at least one-third of the overall food supply.[64] And the UK's Waste and Resources Action Programme ascertained that inedible household food waste (including fruit and vegetable peelings, and bones) is only 30 percent of the total, meaning that 70 percent of wasted food was perfectly edible and was not consumed either because it spoiled or because too much of it was served.[65] Reducing food waste might seem to be much easier than reforming complex production processes, and yet this proverbial low-hanging fruit has been difficult to harvest.

Eliminating waste that takes place all along the long and complex production-processing-distribution-wholesaling-retailing-consumption chain (from fields and barns to plates) is extremely challenging. American food balances show that the nationwide share of wasted food has remained stable during the past 40 years, despite perennial calls for improvements.[66] And higher food waste accompanied China's improving nutrition as the country moved from the precarious food supply that prevailed until the early 1980s to averaging per capita rates that are now higher than in Japan.[67]

Higher food prices should lead to lower waste, but this is not a desirable way to fix the problem in low-income countries—where food access for many disadvantaged families remains precarious and where food still claims a large share of overall family spending—while in affluent nations, where food is relatively inexpensive, this would require substantial price hikes, a policy that has no eager promoters.[68]

In well-off societies, a better way to reduce agriculture's dependence on fossil fuel subsidies is to make appeals for adopting healthy and satisfactory alternatives to today's excessively rich and meaty diets—the easiest choices being moderate meat consumption, and favoring meat that can be grown with lower environmental impact. The quest for mass-scale veganism is doomed to fail. Eating meat has been as significant a component of our evolutionary heritage as our large brains (which evolved partly because of meat eating), bipedalism, and symbolic language.[69] All our hominin ancestors were omnivorous, as are both species of chimpanzees (*Pan troglodytes* and *Pan paniscus*), the hominins closest to us in their genetic makeup; they supplement their plant diet by hunting (and sharing) small monkeys, wild pigs, and tortoises.[70]

Full expression of human growth potential on a population basis can take place only when diets in childhood and adolescence contain sufficient quantities of animal protein, first in milk and later in other dairy products, eggs, and meat: rising post-1950 body heights in Japan, South Korea, and China, as a result of increased intake of animal products, are unmistakable testimonies to this reality.[71] Conversely, most people who become vegetarians or vegans do not remain so for the remainder of their lives. The idea that billions of humans—across the world, not only in affluent Western cities—would willfully not eat any animal products, or that there'd be enough support for governments to enforce that anytime soon, is ridiculous.

But none of this means that we could not eat much *less* meat than affluent countries have averaged during the past two generations.[72] When expressed in terms of carcass weight, annual meat supply in many high-income countries has averaged close to, or even in excess of, 100 kilograms per capita—but the best nutritional advice is that we do not have to eat more than an adult's body mass equivalent in meat per year to obtain an adequate amount of high-quality protein.[73]

While veganism is a waste of valuable biomass (only ruminants—that is cattle, sheep, and goats—can digest such cellulosic plant tissues as straw and stalks), high-level carnivory has no proven nutritional benefits: it certainly does not add any years to life expectancy, and it is a source of additional environmental stress. Meat consumption in Japan, the country with the world's highest longevity, has recently been below 30 kilograms per year; and a much less appreciated fact is that similarly low consumption rates have become fairly common in France, traditionally a nation of high meat intake. By 2013, nearly 40 percent of adult French were *petits consommateurs*, eating meat only in small amounts adding up to less than 39 kg/year, while the heavy meat consumers, averaging about 80 kg/year, made up less than 30 percent of French adults.[74]

Obviously, if all high-income countries were to follow these examples, they could reduce their crop harvests—because most of their grain harvests are not destined directly for food but for animal feed.[75] But this is not a universal option. While meat intakes in many affluent countries have been declining and could be cut even further, they have been rising rapidly in such modernizing nations as Brazil and Indonesia (where they have more than doubled since 1980) and China (where they have quadrupled since 1980).[76] Moreover, there are billions of people in Asia and Africa whose meat consumption remains minimal and whose health would benefit from more meaty diets.

Additional opportunities to reduce the dependence on synthetic nitrogenous fertilizers come on the production side—for example, improving the efficiency of nitrogen uptake by plants. But again, these opportunities are circumscribed. Between 1961 and 1980 there was a substantial decline in the share of applied nitrogen actually incorporated by crops (from 68 percent to 45 percent), then came a levelling off at around 47 percent.[77] And in China, the world's largest consumer of nitrogen fertilizer, only a third of the applied nitrogen is actually used by rice; the rest is lost to the atmosphere and to ground and stream waters.[78] Given that we are expecting at least 2 billion more people by 2050, and that more than twice as many people in the low-income countries of Asia and Africa should see further gains—both in quantity and quality—in their food supply, there is no near-term prospect for substantially reducing the global dependence on synthetic nitrogenous fertilizers.

There are obvious opportunities for running field machinery without fossil fuels. Decarbonized irrigation could become common with pumps powered by solar- or wind-generated electricity rather than by combustion engines. Batteries with improving energy density and lower cost would make it possible to convert more tractors and trucks to electric drive.[79] And in the next chapter I will explain the alternatives to the dominant, natural gas–based synthesis of ammonia. But none of these options can be adopted either rapidly or without additional (and often substantial) investments.

These advances are, at present, a very long way off. They will depend on inexpensive renewable electricity generation backed up by adequate large-scale storage, a combination that is yet to be commercialized (and an alternative to large pumped hydro storage is yet to be invented; for more see chapter 3). A nearly perfect solution would be to develop grain or oil crops with the capabilities common to leguminous plants—that is, with their roots hosting bacteria able to convert inert atmospheric nitrogen to nitrates. Plant scientists have been dreaming about this for decades, but no releases of commercial nitrogen-fixing varieties of wheat or rice are coming anytime soon.[80] Nor is it very likely that all affluent countries and better-off modernizing economies will adopt large-scale voluntary reductions in the quantity and variety of their typical diets, or that the resources (fuel, fertilizers, and machinery) saved by such pullbacks would be transferred to Africa to improve the continent's still-dismal nutrition.

Half a century ago, Howard Odum—in his systematic examination of energy and the environment—noted that modern societies "did not understand the energetics involved and the various means by which the energies entering a complex system are fed back as subsidies indirectly into all parts of the network . . . industrial man no longer eats potatoes made from solar energy; now he eats potatoes partly made of oil."[81]

Fifty years later, this existential dependence is still insufficiently appreciated—but the readers of this book now understand that our food is partly made not just of oil, but also of coal that was used to produce the coke required for smelting the iron needed for field, transportation, and food processing machinery; of natural gas that serves as both feedstock and fuel for the synthesis of nitrogenous fertilizers; and of the electricity generated by the combustion of fossil fuels that is indispensable for crop processing, taking care of animals, and food and feed storage and preparation.

Modern agriculture's higher yields are not produced with a fraction of the labor that was required just a lifetime ago because we have improved the efficiency of photosynthesis, but because we have provided better varieties of crops with better conditions for their growth by supplying them with adequate nutrients and water, by reducing weeds that compete for the same inputs, and by protecting them against pests. Concurrently, our much-increased capture of wild aquatic species has depended on expanding the extent and the intensity of fishing, and the rise of aquaculture could not happen without providing requisite enclosures and high-quality feed.

All these critical interventions have demanded substantial—and rising—inputs of fossil fuels; and even if we try to change the global food system as fast as is realistically conceivable, we will be eating transformed fossil fuels, be it as loaves of bread or as fishes, for decades to come.

3. Understanding Our Material World

The Four Pillars of Modern Civilization

Where it matters, ranking is impossible—or, at least, inadvisable. The heart is not more important than the brain; vitamin C is no less indispensable for human health than vitamin D. Food and energy supply, the two existential necessities covered in the preceding chapters, would be impossible without mass-scale mobilization of many man-made materials—metals, alloys, non-metallic and synthetic compounds—and the same is true about all our buildings and infrastructures and about all modes of transportation and communication. Of course, you would not know this if you were to judge the importance of these materials by the attention they get (or rather do not get), not only from mass media "news" but also from supposedly much more exalted economic analyses or forecasts of notable developments.

All of this coverage deals overwhelmingly with such immaterial, intangible phenomena as the annual percentage growth of GDP (how Western economists used to swoon over China's double-digit rates!), rising national debt ratios (unimportant in the world of Modern Monetary Theory, with money supply seen as unlimited), record sums poured into new initial public offerings (for such existentially critical inventions as gaming apps), the benefits of unprecedented mobile connectivity (awaiting 5G networks as something close to the second coming), or promises of artificial intelligence imminently transforming our lives (the pandemic was an excellent demonstration of the complete emptiness of such claims).

First things first. We could have an accomplished and reasonably affluent civilization that provides plenty of food, material comforts, and access to education and health care, without any semiconductors, microchips, or personal computers: we had one until, respectively, the mid-1950s (first commercial applications of transistors), the early 1970s (Intel's first microprocessors), and the early 1980s (first larger-scale ownership of PCs).[1] And we managed, until the 1990s, to integrate economies, mobilize necessary investments, build requisite infrastructures, and connect the world by wide-body jetliners without any smartphones, social media, and puerile apps. But none of these advances in electronics and telecommunications could have taken place without the assured provision of energies and materials required to embody the inventions in myriads of electricity-consuming components, devices, assemblies, and systems ranging from tiny microprocessors to massive data centers.

Silicon (Si) made into thin wafers (the basic substrate of microchips) is the signature material of the electronic age, but billions of people could live prosperously without it; it is not an existential constraint on modern civilization. Producing large, high-purity (99.999999999 percent pure) silicon crystals that are cut into wafers is a complex, multi-step, and highly energy-intensive process: it costs two orders of magnitude more primary energy than making aluminum from bauxite, and three orders of magnitude more than smelting iron and making steel.[2] But the raw material is super-abundant (Si is the second-most common element in the Earth's crust—nearly 28 percent, compared to 49 percent for oxygen) and the annual output of electronic-grade silicon is very small compared to other indispensable materials, recently on the order of 10,000 tons of wafers.[3]

Of course, annual consumption of a material is not the best indicator of its indispensability, but in this case the verdict is clear: as useful and as transformative as post-1950 electronic advances have been, they do not constitute the indispensable material foundations of modern civilization. And while there can be no indisputable ordering of our material needs based on claims of their importance, I can offer a defensible ranking that considers their indispensability, ubiquity, and the demand size. Four materials rank highest on this combined scale, and they form what I have called the four pillars of modern civilization: cement, steel, plastics, and ammonia.[4]

Physically and chemically, these four materials are distinguished by an enormous diversity of properties and functions. But despite these differences in attributes and specific uses, they share more than their indispensability for the functioning of modern societies. They are needed in larger (and still increasing) quantities than are other essential inputs. In 2019, the world consumed about 4.5 billion tons of cement, 1.8 billion tons of steel, 370 million tons of plastics, and 150 million tons of ammonia, and they are not readily replaceable by other materials—certainly not in the near future or on a global scale.[5]

As noted in chapter 2, only an impossibly complete recycling of all wastes voided by grazing animals could, together with near-perfect recycling of all other sources of organic nitrogen, provide the amount of nitrogen annually applied to crops in ammonia-based fertilizers. Meanwhile, there are no other materials that can rival the combination of malleability, durability, and light weight offered by many kinds of plastics. Similarly, even if we were able to produce identical masses of construction lumber or quarried stone, they could not equal the strength, versatility, and durability of reinforced concrete. We would be able to build pyramids and cathedrals but not elegant long spans of arched bridges, giant hydroelectric dams, multilane roads, or long airport runways. And steel has become so ubiquitous that its irreplaceable deployment determines our ability to extract energies, produce food, and shelter populations, as well as ensuring the extent and quality of all essential infrastructures: no metal could, even remotely, become its substitute.

Another key commonality between these four materials is particularly noteworthy as we contemplate the future without fossil carbon: the mass-scale production of all of them depends heavily on the combustion of fossil fuels, and some of these fuels also supply feedstocks for the synthesis of ammonia and for the production of plastics.[6] Iron ore smelting in blast furnaces requires coke made from coal (and also natural gas); energy for cement production comes mostly from coal dust, petroleum coke, and heavy fuel oil. The vast majority of simple molecules that are bonded in long chains or branches to make plastics are derived from crude oils and natural gases. And in the modern synthesis of ammonia, natural gas is both the source of hydrogen and processing energy.

As a result, global production of these four indispensable materials claims about 17 percent of the world's primary energy supply, and 25 percent of all CO_2 emissions originating in the combustion of fossil fuels—and currently there are no commercially available and readily deployable mass-scale alternatives to displace these established processes.[7] Although there is no shortage of proposals and experimental techniques to produce these materials without relying on fossil carbon—ranging from new catalyses for ammonia synthesis to hydrogen-based steelmaking—none of those alternatives has been commercialized, and even if the aggressive pursuit of non-carbon options were to take place, it would obviously take decades to displace the existing capacities that are producing, at affordable prices, at annual rates of hundreds of millions to billions of tons.[8]

In order to truly appreciate the importance of these materials, I will explain their basic properties and functions, outline briefly the histories of technical advances and epoch-making inventions that made them abundantly available and affordable, and describe the enormous variety of their modern uses. I will start with ammonia—because of its indispensability for feeding a growing share of the global population—and proceed, in the order of annual global mass production, to plastics, steel, and cement.

Ammonia: the gas that feeds the world

Of the four substances (and despite my dislike of rankings!), it is ammonia that deserves the top position as our most important material. As explained in the previous chapter, without its use as the dominant nitrogen fertilizer (directly or as feedstock for the synthesis of other nitrogenous compounds), it would be impossible to feed at least 40 percent and up to 50 percent of today's nearly 8 billion people. Simply restated: in 2020, nearly 4 billion people would not have been alive without synthetic ammonia. No comparable existential constraints apply to plastics or steel, nor to the cement that is required to make concrete (nor, as already noted, to silicon).

Ammonia is a simple inorganic compound of one nitrogen and three hydrogens (NH_3), which means that nitrogen makes up 82 percent of its mass.[9] At atmospheric pressure it is an invisible gas with a characteristic pungent smell of unflushed toilets or decomposing animal manure. Inhaling it in low concentrations causes headaches, nausea, and vomiting; higher concentrations irritate the eyes, nose, mouth, throat, and lungs; and inhalation of very high concentrations can be instantly fatal. In contrast, ammonium (NH_4^+, ammonium ion), formed by the dissolution of ammonia in water, is non-toxic and does not easily penetrate cell membranes.

Synthesizing this simple molecule was surprisingly challenging. The history of inventions includes famous cases of accidental discoveries; in this chapter on materials, the story of Teflon might be the most apposite example. In 1938, Roy Plunkett, a chemist at DuPont, and his assistant Jack Rebok formulated tetrafluoroethylene as a new refrigerant compound. After storing it in refrigerated cylinders, they found that the compound underwent unexpected polymerization, turning into polytetrafluoroethylene, a white, waxy, slippery powder. After the Second World War, Teflon became one of the best-known synthetic materials, and perhaps the only one that made it into political jargon (we have Teflon presidents, but seemingly no Bakelite presidents—though there was an Iron Lady).[10]

The synthesis of ammonia from its elements belongs to the opposite class of discoveries—those with a clearly defined goal pursued by some of the best-qualified scientists and eventually reached by a persevering researcher. The need for this breakthrough was obvious. Between 1850 and 1900 the total population of the industrializing countries of Europe and North America grew from 300 million to 500 million, and rapid urbanization helped to drive a dietary transition from a barely adequate grain-dominated supply to generally higher food energy intakes containing more animal products and sugar.[11] Yields remained stagnant but the dietary shift was supported by an unprecedented expansion of cropland: between 1850 and 1900, about 200 million hectares of North and South American, Russian, and Australian grasslands were converted to grain fields.[12]

Maturing agronomic science made it clear that the only way to secure adequate food for the larger populations of the 20th century was to raise yields by increasing the supply of nitrogen and phosphorus, two key plant macronutrients. The mining of phosphates (first in North Carolina and then in Florida) and their treatment by acids opened the way to a reliable supply of phosphatic fertilizers.[13] But, there was no comparably assured source of nitrogen. The mining of guano (accumulated bird droppings, moderately rich in nitrogen) on dry tropical islands had quickly exhausted the richest deposits, and the rising imports of Chilean nitrates (the country has extensive sodium nitrate layers in its arid northern regions) were insufficient to meet future global demand.[14]

The challenge was to ensure that humanity could secure enough nitrogen to sustain its expanding numbers. The need was explained in 1898 in the clearest possible manner by William Crookes, chemist and physicist, to the British Association for the Advancement of Science, in his presidential address dedicated to the so-called wheat problem. He warned that "all civilized nations stand in deadly peril of not having enough to eat," but he saw the way out: science coming to the rescue, tapping the practically unlimited mass of nitrogen in the atmosphere (present as the unreactive molecule N_2) and converting it into compounds assimilable by plants. He rightly concluded that this challenge "differs materially from other chemical discoveries which are in the air, so to speak, but are not yet matured. The fixation of nitrogen is vital to the progress of civilized humanity. Other discoveries minister to our increased intellectual comfort, luxury, or convenience; they serve to make life easier, to hasten the acquisition of wealth, or to save time, health, or worry. The fixation of nitrogen is a question of the not far-distant future."[15]

Crookes's vision was realized just 10 years after his address. The synthesis of ammonia from its elements, nitrogen and hydrogen, was pursued by a number of highly qualified chemists (including Wilhelm Ostwald, a Nobel Prize winner in chemistry in 1909), but in 1908 Fritz Haber—at that time professor of physical chemistry and electrochemistry at the Technische Hochschule in Karlsruhe—working with his English assistant Robert Le Rossignol and supported by BASF, Germany's (and the world's) leading chemical enterprise, was the first researcher to succeed.[16] His solution relied on using an iron catalyst (a compound that increases the rate of a chemical reaction without altering its own composition) and deploying unprecedented reaction pressure.

It was a no smaller challenge to scale up Haber's experimental success to a commercial enterprise. Under the leadership of Carl Bosch, an expert in chemical as well as metallurgical engineering who joined BASF in 1899, success was achieved in just four years. The world's first ammonia synthesis plant began to operate at Oppau in September 1913, and the term "Haber-Bosch process" has endured ever since.[17]

Within a year, the Oppau plant's ammonia was diverted to make the nitrate needed to produce explosives for the German army. A new, much larger, ammonia factory was completed in 1917 in Leuna, but it did little to prevent Germany's defeat. The postwar expansion of ammonia synthesis proceeded despite the economic crisis of the 1930s, and continued during the Second World War, but by 1950 synthetic ammonia was still far less common than animal manures.[18]

The next two decades saw an eightfold increase of ammonia production to just over 30 million tons a year as synthetic fertilizer enabled the Green Revolution (starting during the 1960s)—the adoption of new superior wheat and rice varieties that, when supplied with adequate nitrogen, produced unprecedented yields. The key innovations behind this rise were the use of natural gas as the source of hydrogen, and the introduction of efficient centrifugal compressors and better catalysts.[19]

Then, as in so many other instances of modern industrial development, post-Mao China took the lead. Mao was responsible for the deadliest famine in history (1958–1961), and when he died in 1976 the country's per capita food supply was hardly better than when he had proclaimed the existence of the communist state in 1949.[20] China's first major business deal to follow President Nixon's trip to Beijing in 1972 was an order for 13 of the world's most advanced ammonia-urea plants from M. W. Kellogg of Texas.[21] By 1984 the country abolished urban food rationing, and by the year 2000 its average daily per capita food supply was higher than Japan's.[22] The only way to make this happen was by breaking the country's nitrogen barrier and raising the annual grain harvest to more than 650 million tons a year.

The best account of recent nitrogen flows in China's agriculture shows that about 60 percent of the nutrient available to the country's crops comes from synthetic ammonia: feeding three out of five of the Chinese population thus depends on the synthesis of this compound.[23] The corresponding global mean is about 50 percent. This dependence easily justifies calling the Haber-Bosch synthesis of ammonia perhaps the most momentous technical advance in history. Other inventions, as William Crookes correctly judged, minister to our comforts, convenience, luxury, wealth, or productivity, and others yet save our lives from premature death and chronic disease—but without the synthesis of ammonia, we could not ensure the very survival of large shares of today's and tomorrow's population.[24]

I hasten to add that the 50 percent of humanity dependent on ammonia is not an immutable approximation. Given prevailing diets and farming practices, synthetic nitrogen feeds half of humanity—or, everything else being equal, half of the world's population could not be sustained without synthetic nitrogenous fertilizers. But the share would be lower if the affluent world converted to the largely meatless Indian diet, and it would be higher if the entire world ate as well as the Chinese do today, to say nothing about the universal adoption of the American diet.[25] We could also reduce our dependence on nitrogenous fertilizers by cutting our food waste (as we saw earlier) and by using the fertilizers more efficiently.

About 80 percent of global ammonia production is used to fertilize crops; the rest is used to make nitric acid, explosives, rocket propellants, dyes, fibers, and window and floor cleaners.[26] With proper precautions and special equipment, ammonia can be applied directly to fields;[27] but the compound is mostly used as the indispensable feedstock for producing solid and liquid nitrogenous fertilizers. Urea, the solid fertilizer with the highest nitrogen content (46 percent), dominates.[28] Recently, it has accounted for about 55 percent of all nitrogen applied to the world's fields, and it is widely used in Asia to support the rice and wheat harvests of China and India—the world's two most populous nations—and to guarantee good yields in five other Asian countries with more than 100 million inhabitants.[29]

Less important nitrogenous fertilizers include ammonium nitrate, ammonium sulfate, and calcium ammonium nitrate, as well as various liquid solutions. Once the nitrogen fertilizers are applied to fields, it is almost impossible to control their natural losses due to volatilization (from ammonia compounds), leaching (nitrates are readily soluble in water), and denitrification (the bacteria-driven conversion of nitrates back to nitrogen molecules in the air).[30]

There are now only two effective direct solutions to field losses of nitrogen: the spreading of expensive slow-release compounds; and, more practically, turning to precision farming and applying fertilizers only as needed based on analyses of the soil.[31] As already noted, indirect measures—including higher food prices and reduced meat consumption—could be effective but are not highly popular. As a result, it is unlikely that any realistically conceivable combination of these solutions can bring about a radical change to the global consumption of nitrogenous fertilizers. About 150 megatons of ammonia are now synthesized annually, with about 80 percent used as fertilizer. Nearly 60 percent of that fertilizer is applied in Asia, about a quarter in Europe and North America, and less than 5 percent in Africa.[32] Most rich countries certainly could, and should, pull back on their high application rates (their average per capita food supply is already too high), and China and India—two heavy users—have many opportunities to reduce their excessive fertilizer applications.

But Africa, the continent with the fastest-growing population, remains deprived of the nutrient and is a substantial food importer. Any hope for its greater food self-sufficiency rests on the increased use of nitrogen: after all, the continent's recent usage of ammonia has been less than a third of the European mean.[33] The best (and long-sought) solution to boost nitrogen supply would be to endow non-leguminous plants with nitrogen-fixing capabilities, a promise genetic engineering is yet to deliver on, while a less radical option—inoculating seeds with a nitrogen-fixing bacterium—is a recent innovation whose eventual commercial extent is still unclear.

Plastics: diverse, useful, troublesome

Plastics are a large group of synthetic (or semisynthetic) organic materials whose common quality is that they are fit for forming (molding). Synthesis of plastics begins with monomers, simple molecules that can be bonded in long chains or branches to make polymers. The two key monomers, ethylene and propylene, are produced by the steam cracking (heating to 750–950°C) of hydrocarbon feedstocks, and hydrocarbons also energize subsequent syntheses.[34] The malleability of plastics makes it possible to form them by casting, pressing, or extruding, and they create shapes ranging from thin film to heavy-duty pipes and from feather-light bottles to massive and sturdy waste containers.

Global output has been dominated by thermoplastics—polymers that soften readily when heated and harden again when cooled. Low- and high-density polyethylene (PE) now accounts for more than 20 percent of the world's plastic polymers, polypropylene (PP) for about 15 percent, and polyvinyl chloride (PVC) for more than 10 percent.[35] In contrast, thermoset plastics (including polyurethanes, polyimides, melamine, and urea-formaldehyde) resist softening when heated.

Some thermoplastics combine low specific gravity (light weight) with fairly high hardness (durability). Durable aluminum weighs only a third as much as carbon steel, but PVC density is less than 20 percent and PP less than 12 percent compared to steel; and while the ultimate tensile strength of structural steel is 400 megapascals, that of polystyrene is, at 100 megapascals, twice that of wood or glass and only 10 percent less than that of aluminum.[36]

This combination of low weight and high strength has made thermoplastics the preferred choice for such applications as heavy-duty pipes and flanges, anti-skid surfaces, and chemical tanks. Thermoplastic polymers have found widespread uses in car interiors and exteriors (PP bumpers, PVC dashboards and car parts, polycarbonate headlight lenses); light high-temperature or flame-retardant thermoplastics (polycarbonate, PVC/acrylic blends) dominate the interiors of modern aircraft; and carbon-fiber reinforced plastics (composite materials) are now used for building aircraft airframes.[37]

The first plastics, most notably celluloid made from cellulose nitrate and camphor (later the highly flammable mainstay of the film industry, displaced only during the 1950s), were produced in small amounts during the last three decades of the 19th century, but the first thermoset material (molded at 150–160°C) was prepared in 1907 by Leo Hendrik Baekeland, a Belgian chemist working in New York.[38] His

General Bakelite Company, founded in 1910, was the first industrial producer of a plastic that was molded into pieces ranging from electric insulators to black rotary dial telephones and, during the Second World War, used for parts in lightweight weapons. Meanwhile, cellophane was invented in 1908 by Jacques Brandenberger.

During the interwar years came the first large-scale syntheses of PVC, which was discovered as early as 1838 but never used outside a laboratory, and DuPont in the US, Imperial Chemical Industries (ICI) in the UK, and IG Farben in Germany funded (very successfully) research dedicated to the discovery of new plastic materials.[39] Before the Second World War this resulted in the commercial production of cellulose acetate (now in absorbent cloths and wipes), neoprene (synthetic rubber), polyester (for fabrics and upholstery), polymethyl methacrylate (otherwise known as plexiglass, and now even more commonly used thanks to the COVID-induced resurgence of dividers and shields). Nylon has been produced since 1938 (toothbrush bristles and stockings were the first commercial products; it now goes into products ranging from fishing nets to parachutes) and—as already noted—so has Teflon, a ubiquitous nonstick coating. The affordable production of styrene also began during the 1930s, and the material is now mostly used as polystyrene, or PS, in packing materials and disposable cups and plates.

IG Farben introduced polyurethanes in 1937 (furniture foams, insulation); ICI used very high pressure to synthesize polyethylene (used in packaging and insulation) and began producing methyl methacrylate (for adhesives, coatings, and paints) in 1933. Polyethylene terephthalate (PET)—since the 1970s the scourge of the planet in the form of discarded drink bottles—was patented in 1941 and mass-produced since the early 1950s (the infernal PET bottle was patented in 1973).[40] The best-known post–Second World War additions include polycarbonates (for optical lenses, windows, rigid covers), polyimide (for medical tubing), liquid crystal polymers (above all for electronics), and such famous DuPont trademarks as Tyvek® (1955), Lycra® (1959), and Kevlar® (1971).[41] By the end of the 20th century, 50 different kinds of plastics were on the global market, and this new diversity—together with rising demand for the most commonly used compounds (PE, PP, PVC, and PET)—led to exponential growth in demand.

Global production rose from only about 20,000 tons in 1925 to 2 million tons by 1950, 150 million tons by the year 2000, and about 370 million tons by 2019.[42] The best way to appreciate the ubiquity of plastic materials in our daily lives is to note how many times a day our hands touch, our eyes see, our bodies rest on, and our feet tread on a plastic: you might be astonished at the frequency of such encounters! As I am typing this: the keys of my Dell laptop and a wireless mouse under my right palm are made of acrylonitrile butadiene styrene, I sit on a swivel chair upholstered in a polyester fabric, and its nylon wheels rest on a polycarbonate carpet protection mat that covers a polyester carpet . . .

An industry that started by supplying small industrial parts (a gear-lever knob in the 1916 Rolls-Royce was the first application) and various household items has vastly expanded those two original commercial niches (most prominently with consumer electronics adding billions of new plastic-dependent items every year) and has added mass-scale applications ranging from car bodies and complete interiors of airplanes to large-diameter pipes.

But plastics have found their most indispensable roles in health care in general and in the hospital treatment of infectious diseases in particular. Modern life now begins (in maternity wards) and ends (in intensive care units) surrounded by plastic items.[43] And those people who had no prior understanding of plastics' role in modern health care got their lesson thanks to COVID-19. The pandemic has taught us this in often drastic ways, as doctors and nurses in North America and Europe ran out of personal protective equipment (PPE)—disposable gloves, masks, shields, hats, gowns, and booties—and as governments outbid each other in order to airlift limited (and highly overpriced) supplies from China, to which the Western producers of PPE, obsessed with cutting costs, had relocated most of their production lines, creating dangerous yet entirely avoidable supply shortages.[44]

Plastic items in hospitals are made above all from different kinds of PVC: flexible tubes (used for feeding patients, delivering oxygen, and monitoring blood pressure), catheters, intravenous containers, blood bags, sterile packaging, assorted trays and basins, bedpans and bed rails, thermal blankets, and countless pieces of labware. PVC is now the primary component in more than a quarter of all health-care products, and in modern homes it is present in wall and roof membranes, window frames, blinds, hoses, cable insulation, electronic components, a still-growing array of office supplies, and toys—and as credit cards used to purchase all of the above.[45]

Recent years have seen rising concerns about plastic pollution on land and even more so in the ocean, coastal waters, and on beaches. I will return to this in the environmental chapter, but this irresponsible dumping of plastics is not an argument against the proper use of these diverse and often truly indispensable synthetic materials. Moreover, as far as microfibers go, it is wrong to assume, as so many do, that most of their presence in ocean water derives from the wear and tear of synthetic textiles. Those polymers now account for two-thirds of global fiber output, but a study of seawater samples showed that oceanic fibers are mainly (>90 percent) of natural origin.[46]

Steel: ubiquitous and recyclable

Steels (the plural is more accurate as there are more than 3,500 varieties) are alloys dominated by iron (Fe).[47] Pig or cast iron, the hot metal produced by blast furnaces, is typically 95–97 percent iron, 1.8–4 percent carbon, and 0.5–3 percent silicon, with mere traces of a few other elements.[48] Its high carbon content makes it brittle, it has low ductility (the ability to stretch), and its tensile strength (resistance to breaking under tension) is inferior to that of bronze or brass. Preindustrial steel was made in Asia and in Europe by a variety of artisanal methods—that is, always laboriously and expensively—and hence it was never available for common use.[49]

Modern steels are made from cast iron by reducing its high carbon content to 0.08–2.1 percent by weight. Steel's physical properties handily beat those of the hardest stones, as well as those of the other two most common metals. Granite has a similar compressive strength (capacity to withstand loads that shorten the material) but its tensile strength is an order of magnitude lower: granite columns bear their load as well as steel, but steel beams can bear loads 15–30 times higher.[50] Steel's typical tensile strength is about seven times that of aluminum and nearly four times that of copper; its hardness is, respectively, four and eight times higher; and it is heat resistant—aluminum melts at 660°C, copper at 1,085°C, steel only at 1,425°C.

Steels come in four major categories.[51] Carbon steels (90 percent of all steels on the market are 0.3–0.95 percent carbon) are everywhere, from bridges to fridges and from gears to shears. Alloy steels include varying shares of one or more elements (most commonly manganese, nickel, silicon, and chromium, but also aluminum, molybdenum, titanium, and vanadium), added in order to improve their physical properties (hardness, strength, ductility). Stainless steel (10–20 percent chromium) was made for the first time only in 1912 for kitchenware, and is now widely used for surgical instruments, engines, machine parts, and in construction.[52] Tool steels have a tensile strength 2–4 times higher than the best construction steels, and they are used for cutting steel and other metals for dies (for stamping or extrusion of other metals or plastics), as well as for manual cutting and hammering. And all steels (except for some stainless varieties) are magnetic and hence suitable for making electric machinery.

Steel determines the look of modern civilization and enables its most fundamental functions. This is the most widely used metal and it forms countless visible and invisible critical components of today's world. Moreover, nearly all other metallic and non-metallic products we use have been extracted, processed, shaped, finished, and distributed with tools and machines made of steel, and no mode of today's mass transportation could function without steel. Naked steel is ubiquitous inside and outside of our homes, in items small (cutlery, knives, cooking pots, pans, kitchen gadgets, garden tools) and large (appliances, lawnmowers, bicycles, cars).

Before large city buildings go up, you can see massive steel pile-driving machines ramming in steel or steel-reinforced concrete for the foundations, and then the site is dominated for months by tall steel construction cranes. In 1954, New York's Socony-Mobil Building was the first skyscraper entirely clad in stainless steel, and more recently Dubai's 828-meter-tall Burj Khalifa uses textured stainless-steel spandrel panels and vertical tubular steel pins.[53] Steel is both the critical structural component and the design feature of many elegant cantilevered and suspension bridges:[54] in San Francisco's Golden Gate Bridge it is constantly repainted orange;[55] Japan's Akashi Kaikyō Bridge has the world's longest central span, at nearly 2 kilometers, and its steel towers support woven steel cables that are 1.12 meters in diameter.[56]

City streets are lined by regularly spaced lighting poles made from hot-dip galvanized and powder-coated steel for rust resistance; rolled steel makes roadside traffic signs and structures for overhead signage; and corrugated steel is used for crash barriers. Steel towers support thick steel wires to lift downhill skiers by the millions and to carry visitors in cable cars to tall peaks. Radio and TV towers (guyed masts) broke many height records for man-made structures, and modern landscapes contain seemingly endless repetitions of high-voltage electricity transmission towers. Two recent, prominent additions are dizzyingly tall guyed towers (to carry mobile telephone signals) and groups of large wind turbine towers, both onshore and offshore; and the most massive steel assemblies in the ocean are huge oil and gas production platforms.[57]

By weight, steel is nearly always the largest part of transportation equipment. Jetliners are a major exception (aluminum alloys and composite fibers dominate) and steel is about 10 percent of the plane's weight, in engines and the landing gear.[58] The average car contains about 900 kilograms of steel.[59] With nearly 100 million motor vehicles produced per year, this translates to about 90 million tons of the metal—about 60 percent of it being high-strength steel that makes vehicles 26–40 percent lighter than conventional steel.[60] Although modern high-speed trains (aluminum bodies, plastic interiors) are only about 15 percent steel (wheels, axles, bearings, and motors), their operation requires dedicated tracks using heavier-than-normal steel rails.[61]

Ship hulls of oil and liquefied-gas tankers and of bulk carriers transporting ores, grains, or cement are made by bending large plates of high-tensile steel into desired shapes and welding them together. But the greatest revolution in postwar shipping has been the deployment of container ships (for more detail, see chapter 4). They transport cargo in steel crates of standardized dimensions.[62] These steel boxes are about 2.5 meters high and wide (length varies) and are stacked inside the hulls and high above the deck. Chances are that everything you wear was carried to its final point of sale in a steel container that started its journey in a factory in Asia.

And how were all these tools and machines made? Mostly by other machines and assemblies made largely of steel and which do casting, forging, rolling, subtractive machining (turning, milling, hollowing, and drilling), bending, welding, sharpening, and cutting, the latter operations being possible thanks to amazing tool steels that cut carbon steels as easily as a knife goes through soft butter. And the machines that make machines are mostly powered by electricity, whose generation (and hence also the entire universe of electronics, computing, and telecommunication) is impossible without steel: tall boilers crammed with steel tubes and filled with pressurized water; nuclear reactors enclosed in thick pressure vessels; expanding steam rotating large turbines whose long shafts are machined from rough, massive steel forgings.

Steel that is out of sight underground includes fixed and moving props in deep mines, and millions of kilometers of exploratory, casing, and production pipes in crude oil and natural gas wells. The oil and gas industry also depends on steel buried close to the surface (1–2 meters deep) in gathering, transmission, and distribution pipelines. Trunk lines use pipes with diameters of more than 1 meter, while distribution gas lines may be just 5 centimeters across.[63] Crude oil refineries are essentially forests of steel, with tall distillation columns, catalytic crackers, extensive piping, and storage vessels. Finally, I must note how steel saves lives in hospitals (from centrifuges and diagnostic machines to stainless-steel scalpels, surgical hooks, and retractors), and how it also kills: armies and fleets with their vast arrays of weapons are nothing but enormous repositories of steel dedicated to destruction.[64]

Can we secure the required massive supply of steel and how consequential is the metal's global production? Do we have adequate supplies of iron ore to keep making steel for many generations to come? Can we produce enough of it in order to build modern infrastructures and raise the standards of living in low-income countries, where average per capita steel consumption is even lower than it was in the affluent economies a century ago? Is steelmaking environmentally friendly or is it exceptionally damaging? Can we produce the metal without using any fossil fuels?

The answer to the second question is unequivocally positive. Iron is the Earth's dominant element by mass because it is heavy (nearly eight times as heavy as water) and because it forms the planet's core.[65] But it is also abundant in the Earth's crust: only three elements (oxygen, silicon, and aluminum) are more common; iron, with nearly 6 percent, ranks fourth.[66] Annual production of iron ore—led by Australia, Brazil, and China—is now about 2.5 billion tons; world resources are in excess of 800 billion tons, containing nearly 250 billion tons of the metal. This is a resource/production (R/P) ratio of more than 300 years, far beyond any conceivable planning horizons (the R/P ratio for crude oil is just 50 years).[67]

Moreover, steel is readily recycled by melting it in an electric arc furnace (EAF)—a massive cylindrical heat-resistant container made of heavy steel plates (lined with magnesium bricks), with a removable dome-like water-cooled lid through which three massive carbon electrodes are inserted. After loading the steel scrap, the electrodes are lowered into it, and electric current passing through them forms an arc whose high temperature (1,800°C) easily melts the charged metal.[68] However, their electricity demand is enormous: even a highly efficient modern EAF needs as much electricity every day as an American city of about 150,000 people.[69]

Vehicle recycling is preceded by draining all fluids, ripping out upholstery, and removing batteries, servomotors, tires, radios, and working engines, as well as plastic, rubber, glass, and aluminum components. Car crushers then flatten the stripped bodies preparatory to shredding. By far the most challenging recycling operation is the dismantling of large ocean-going vessels, done mostly on beaches in Pakistan (Gadani, northwest of Karachi), India (Alang in Gujarat), and Bangladesh (near Chittagong). Stripped hulls made of heavy steel plates must be cut by gas and plasma torches—dangerous and polluting work done too often by men working without proper protective gear.[70]

Affluent economies now recycle nearly all of their automotive scrap, have a similarly high rate (>90 percent) for reusing structural steel beams and plates, and only a slightly lower rate for recycling household appliances, and the US has recently recycled more than 65 percent of reinforcement bars in concrete, a rate similar to the recycling of beverage and food steel cans.[71] Steel scrap has become one of the world's most valuable export commodities, as countries with a long history of steel production and with plenty of accumulated scrap sell the material to expanding producers. The EU is the largest exporter, followed by Japan, Russia, and Canada; and China, India, and Turkey are the top buyers.[72] Recycled steel accounts for almost 30 percent of the metal's total annual output, with national shares ranging from 100 percent for several small steel producers to almost 70 percent in the US, about 40 percent in the EU, and to less than 12 percent in China.[73]

This means that primary steelmaking still dominates, producing more than twice as much hot metal every year as is recycled—almost 1.3 billion tons in 2019. The process starts with blast furnaces (tall iron and steel structures lined with heat-resistant materials) that produce liquid (cast or pig) iron by smelting iron ore, coke, and limestone.[74] The second step—reducing cast iron's high carbon content and producing steel—takes place in a BOF (basic oxygen furnace; the adjective refers to the chemical properties of the produced slag). The process was invented during the 1940s and was rapidly commercialized after the mid-1950s.[75] Today's BOFs are large, pear-shaped vessels with an open top used to charge up to 300 tons of hot iron, which gets blasted with oxygen blown in from both top and bottom. The reaction reduces the metal's carbon content (to as little as 0.04 percent) in about 30 minutes. The combination of a blast furnace and a basic oxygen furnace is the basis of modern integrated steelmaking. Final steps include the transfer of hot steel to continuous casting machines to produce steel slabs, billets (square or rectangular shapes), and strips that are eventually converted into final steel products.

Ironmaking is highly energy-intensive, with about 75 percent of the total demand claimed by blast furnaces. Today's best practices have a combined demand of just 17–20 gigajoules per ton of finished product; less efficient operations require 25–30 GJ/t.[76] Obviously, the energy cost of secondary steel made in EAFs is much lower than the cost of integrated production: today's best performance is just above 2 GJ/t. To this must be added the energy costs of rolling the metal (mostly 1.5–2 GJ/t), and hence the representative global rates for the overall energy cost may be about 25 GJ/t for integrated steelmaking and 5 GJ/t for recycled steel.[77] The total energy requirement of global steel production in 2019 was about 34 exajoules, or about 6 percent of the world's primary energy supply.

Given the industry's dependence on coking coal and natural gas, steelmaking has been also a major contributor to the anthropogenic generation of greenhouse gases. The World Steel Association puts the average global rate at 500 kilograms of carbon per ton, with recent primary steelmaking emitting about 900 megatons of carbon a year, or 7–9 percent of direct emissions from the global combustion of fossil fuels.[78] But steel is not the only major material responsible for a significant share of CO_2 emissions: cement is much less energy-intensive, but because its global output is nearly three times that of steel, its production is responsible for a very similar share (about 8 percent) of emitted carbon.

Concrete: a world created by cement

Cement is the indispensable component of concrete, and it is produced by heating (to at least 1,450°C) ground limestone (a source of calcium) and clay, shale, or waste materials (sources of silicon, aluminum, and iron) in large kilns—long (100–220 meters) inclined metal cylinders.[79] This high-temperature sintering produces clinker (fused limestone and aluminosilicates) that is ground to yield fine, powdery cement.

Concrete consists largely (65–85 percent) of aggregates and also water (15–20 percent).[80] Finer aggregates such as sand result in stronger concrete, but need more water in the mix than do coarser aggregates that use different sizes of gravel. The mixture is held together by cement—typically 10–15 percent of concrete's final mass—whose reaction with water first sets the mixture and then hardens it.

The result is now the most massively deployed material of modern civilization, hard and heavy and able to withstand decades of punishing use, particularly when it is reinforced with steel. Plain concrete is fairly good in compression (and the best modern varieties are five times stronger than those of two generations ago)—but weak in tension.[81] Structural steel has tension strength up to 100 times higher, and different types of reinforcing (steel mesh, steel bars, glass or steel fibers, PP) have been used to narrow this huge gap.

Since 2007, most of humanity has lived in cities made possible by concrete. Of course, there are plenty of other materials in urban buildings: skyscrapers have steel skeletons covered by glass or metal; detached houses in North American suburbs are made of wood (studs, plywood, particle board) and gypsum drywall (and are often sheathed in brick or stone); and engineered lumber is now used to build apartments many stories high.[82] But skyscrapers and tall apartment buildings stand on concrete piles, concrete goes not only into foundations and basements but also into many walls and ceilings, and it is ubiquitous in all urban infrastructures—from buried engineering networks (large pipes, cable channels, sewers, subway foundations, tunnels) to aboveground transportation infrastructure (sidewalks, roads, bridges, shipping piers, airport runways). Modern cities—from São Paulo and Hong Kong (with their multistoried apartment towers) to Los Angeles and Beijing (with their extensive networks of freeways)—are embodiments of concrete.

Roman cement was a mixture of gypsum, quicklime, and volcanic sand, and it proved to be an excellent and durable material for large structures, including expansive vaults. The Pantheon, intact after nearly two millennia (it was completed in 126 CE) still spans a greater distance than any other structure made of non-reinforced concrete.[83] But the preparation of modern cement was patented only in 1824 by Joseph Aspdin, an English bricklayer. His hydraulic mortar was made by firing limestone and clay at high temperatures: lime, silica, and alumina present in these materials are vitrified or transformed into a glass-like substance, whose grinding produced Portland cement.[84] Aspdin chose that name (still widely used today) because once hardened, and after reacting with water, the glassy clinker had a color similar to limestone from the Isle of Portland in the English Channel.

As already noted, the new material was excellent in compression, and today's best concretes can withstand pressure of more than 100 megapascals, which is about the weight of an African bull elephant balanced on a coin.[85] Tension is a different matter: a pulling force of just 2 to 5 megapascals (less than it takes to tear human skin) can break concrete apart. That is why the large-scale commercial adoption of concrete in construction took place only after gradual advances in steel reinforcement made it suitable for structural parts subject to high tension.

During the 1860s and 1870s, the first reinforcing patents were filed by François Coignet and Joseph Monier in France (Monier, a gardener, began to use iron mesh to reinforce his planters), but the real breakthrough came in 1884 with Ernest Ransome's reinforcing steel bars.[86] The earliest designs of modern cement rotary kilns, where the minerals are vitrified at temperatures of up to 1,500°C, appeared during the 1890s and made it possible to use affordable concrete in large projects. The sixteen-story Ingalls Building in Cincinnati became the world's first reinforced concrete skyscraper in 1903.[87] Just three years later, Thomas Edison became convinced that concrete should replace wood in the building of American detached houses, and began to design and cast in place concrete homes in New Jersey; in 1911 he tried to revive the failed project by also offering cheap concrete furniture, including entire bedroom sets, and even made a phonograph, one of his favorite inventions, out of concrete.[88]

At the same time, in contrast to Edison's failure, Robert Maillart, a Swiss engineer, pioneered a concrete construction trend that is still going strong: reinforced concrete bridges, starting with the relatively short Zuoz in 1901 and Tavanasa in 1906. His most famous design, the bold Salginatobel arch above an Alpine ravine, was completed in 1930 and is now an International Historic Civil Engineering Landmark.[89] Early concrete designs were also favored by architects Auguste Perret in France (elegant apartments and the Théâtre des Champs-Élysées) and Frank Lloyd Wright in the US. Wright's most famous interwar concrete designs were Tokyo's Imperial Hotel, finished just before the 1923 earthquake leveled the city and damaged the new structure, and Fallingwater in Pennsylvania, completed in 1939. The Guggenheim Museum in New York was his last famous concrete design, completed in 1959.[90]

The tensile strength of reinforcing steel was further improved by pouring concrete into forms whose wires or bars were tensioned just before pouring the concrete (pre-stressing, with end anchors that are used to tension the steel and are released once the concrete bonds with the metal) or after it (post-stressing, with steel tendons locked in place inside protective sleeves). The first major pre-stressed design, Eugène Freyssinet's Plougastel Bridge near Brest, was finished in 1930.[91] With its bold, white, sail-like design, Jørn Utzon's Sydney Opera House (built between 1959 and 1973) is perhaps the world's most famous pre-stressed concrete strucuture.[92] Pre-stressing is now common, and the longest reinforced concrete bridges are not crossing rivers or ravines but rather railroad viaducts for high-speed trains. The record goes to the 164.8-kilometer Danyang–Kunshan Grand Bridge in China (completed in 2010), part of the Beijing–Shanghai high-speed railway.[93]

Reinforced concrete is now inside every large modern building and every transportation infrastructure, from port piers to segmental rings installed by modern tunnel-boring machines (under the Channel and the Alps). The standard configuration of the US Interstate Highway System is a layer of about 28 centimeters of non-reinforced concrete on top of a twice-as-thick layer of natural aggregates (stones, gravel, sand)—and the entire Interstate system contains about 50 million tons of cement, 1.5 billion metric tons of aggregates, and only about 6 million tons of steel (for structural support and culvert pipes).[94] Airport runways (up to 3.5 kilometers long) have reinforced concrete foundations, deepest (up to 1.5 meters) in the touchdown zone to handle the repeated pounding of hundreds of thousands of landings every year by airplanes weighing up to about 380 tons (the Airbus 380). For example, Canada's longest runway (4.27 kilometers, in Calgary) required more than 85,000 cubic meters of concrete and 16,000 tons of reinforcing steel.[95]

But by far the most massive structures built of reinforced concrete are the world's largest dams. The era of these megastructures began during the 1930s with the construction of the Hoover Dam on the Colorado River and the Grand Coulee Dam on the Columbia River. The vertiginous Hoover Dam, located in a gorge southeast of Las Vegas, required about 3.4 million cubic meters of concrete and 20,000 tons of reinforcing steel, twice as much plate and pipe steel, and 8,000 tons of structural steel.[96] Hundreds of these massive structures were built during the second half of the 20th century, and the world's largest dam—China's Sanxia (Three Gorges) on the Yangzi, generating electricity since 2011—has its almost 28 million cubic meters of concrete reinforced with 256,500 tons of steel.[97]

Annual American cement consumption rose tenfold between 1900 and 1928, when it reached 30 million tons, and the postwar construction boom—including the construction of the Interstate Highway System and the expansion of the country's airports—tripled by the century's end. The peak was reached at about 128 million tons in 2005, and recent rates are around 100 million tons a year.[98] This is now a tiny fraction of the annual demand in the world's number-one cement consumer, China. In 1980, at the outset of its modernization drive, it produced less than 80 million tons of cement. In 1985, it surpassed the US to become the world's largest producer, and in 2019 its output of about 2.2 billion tons was just over half of the global total.[99]

Perhaps the most stunning outcome of this rise is that in just two years—2018 and 2019—China produced nearly as much cement (about 4.4 billion tons) as did the United States during the entire 20th century (4.56 billion tons). Not surprisingly, the country now has the world's most extensive systems of freeways, rapid trains, and airports, as well as the largest number of giant hydro stations and new multimillion-population cities. Yet another astounding statistic is that the world now consumes in one year more cement than it did during the entire first half of the 20th century. And (both fortunately and unfortunately) these enormous masses of modern concrete will not last as long as the Pantheon's coffered dome.

Ordinary construction concrete is not a highly durable material and it is subject to many environmental assaults.[100] Exposed surfaces are attacked by moisture, freezing, bacterial and algal growth (especially in the tropics), acid deposition, and vibration. Buried concrete structures suffer from crack-inducing pressures, and from damage caused by reactive compounds seeping from above. Concrete's high alkalinity (freshly poured material has a pH of about 12.5) is an effective guard against the corrosion of the reinforcing steel, but cracks and exfoliation expose the metal to corrosive disintegration. Chlorides attack concrete submerged in seawater and concrete on winter roads where salt is used for de-icing.

Between 1990 and 2020, the mass-scale concretization of the modern world has emplaced nearly 700 billion tons of hard but slowly crumbling material. The durability of concrete structures varies widely: while it is impossible to offer an average longevity figure, many will deteriorate badly after just two or three decades while others will do well for 60–100 years. This means that during the 21st century we will face unprecedented burdens of concrete deterioration, renewal, and removal (with, obviously, a particularly acute problem in China), as structures will have to be torn down—in order to be replaced or destroyed—or abandoned. Concrete structures can be slowly demolished, reinforcing steel can be separated, and both materials can be recycled: not cheap, but perfectly possible. After crushing and sieving, the aggregate can be incorporated in new concrete, and reinforcing steel can be recycled.[101] Even now, replacement concrete and new concrete are needed everywhere.

In affluent countries with low population growth, the main need is to fix decaying infrastructures. The latest report card for the US awards nothing but poor to very poor grades to all sectors where concrete dominates, with dams, roads, and aviation getting Ds and the overall average grade just D+.[102] This appraisal gives an inkling of what China might face (mass- and money-wise) by 2050. In contrast, the poorest countries need essential infrastructures and the most basic need in many homes in Africa and Asia is to replace mud floors with concrete floors in order to improve overall hygiene and to reduce the incidence of parasitic diseases by nearly 80 percent.[103]

With aging populations, migration to cities, economic globalization, and widespread regional declines, more concrete will simply be abandoned worldwide. Concrete ruins of car factories in Detroit, abandoned enterprises in Europe's old industrial regions, and all those now derelict plants and monuments built by Soviet central planners on the Russian plain and in Siberia are only the first waves of this trend.[104] Other common and prominent concrete relics are thick-walled defensive bunkers such as those of Normandy and the Maginot Line, and massive concrete silos that formerly housed nuclear missiles and now sit empty on America's Great Plains.

Material outlook: old and new inputs

During the first half of the 21st century—with slower global population growth and with stagnant or even declining counts in many affluent countries—economies should have no problems meeting the demand for steel, cement, ammonia, and plastics, especially with intensified recycling. But it is unlikely that by 2050 all of these industries will eliminate their dependence on fossil fuels and cease to be significant contributors to global CO_2 emissions. This is especially unlikely in today's low-income modernizing countries, whose enormous infrastructural and consumer needs will require large-scale increases of all basic materials.

Replicating the post-1990 Chinese experience in those countries would amount to a 15-fold increase of steel output, a more than 10-fold boost for cement production, a more than doubling of ammonia synthesis, and a more than 30-fold increase of plastic syntheses.[105] Obviously, even if other modernizing countries accomplish only half or even just a quarter of China's recent material advances, these countries would still see multiplications of their current uses. Requirements for fossil carbon have been—and for decades will continue to be—the price we pay for the multitude of benefits arising from our reliance on steel, cement, ammonia, and plastics. And as we continue to expand renewable energy conversions, we will require larger masses of old materials as well as unprecedented quantities of materials that were previously needed in only modest amounts.[106]

Two prominent examples illustrate this unfolding material dependence. No structures are more obvious symbols of "green" electricity generation than large wind turbines—but these enormous accumulations of steel, cement, and plastics are also embodiments of fossil fuels.[107] Their foundations are reinforced concrete, their towers, nacelles, and rotors are steel (altogether nearly 200 tons of it for every megawatt of installed generating capacity), and their massive blades are energy-intensive—and difficult to recycle—plastic resins (about 15 tons of them for a midsize turbine). All of these giant parts must be brought to the installation sites by outsized trucks and erected by large steel cranes, and turbine gearboxes must be repeatedly lubricated with oil. Multiplying these requirements by the millions of turbines that would be needed to eliminate electricity generated from fossil fuels shows how misleading any talks are about the coming dematerialization of green economies.

Electric cars provide perhaps the best example of new, and enormous, material dependencies. A typical lithium car battery weighing about 450 kilograms contains about 11 kilograms of lithium, nearly 14 kilograms of cobalt, 27 kilograms of nickel, more than 40 kilograms of copper, and 50 kilograms of graphite—as well as about 181 kilograms of steel, aluminum, and plastics. Supplying these materials for a single vehicle requires processing about 40 tons of ores, and given the low concentration of many elements in their ores it necessitates extracting and processing about 225 tons of raw materials.[108] Again, we would have to multiply this by close to 100 million units, which is the annual worldwide production of internal-combustion vehicles that would have to be replaced by electric drive.

Uncertainties about the future rates of electric vehicle adoption are large, but a detailed assessment of material needs, based on two scenarios (assuming that 25 percent or 50 percent of the global fleet in 2050 would be electric vehicles), found the following: from 2020 to 2050 demand for lithium would grow by factors of 18–20, for cobalt by 17–19, for nickel by 28–31, and factors of 15–20 would apply for most other materials from 2020.[109] Obviously, this would require not only a drastic expansion of lithium, cobalt (a large share of it now coming from Congo's perilously hand-dug deep shafts and from widespread child labor), and nickel extraction and processing, but also an extensive search for new resources. And these, in turn, could not take place without large additional conversions of fossil fuels and electricity. Generating smoothly rising forecasts of future electric vehicle ownership is one thing; creating these new material supplies on a mass global scale is quite another.

Modern economies will always be tied to massive material flows, whether those of ammonia-based fertilizers to feed the still-growing global population; plastics, steel, and cement needed for new tools, machines, structures, and infrastructures; or new inputs required to produce solar cells, wind turbines, electric cars, and storage batteries. And until all energies used to extract and process these materials come from renewable conversions, modern civilization will remain fundamentally dependent on the fossil fuels used in the production of these indispensable materials. No AI, no apps, and no electronic messages will change that.

4. Understanding Globalization

Engines, Microchips, and Beyond

Globalization manifests itself in countless quotidian ways. Ships loaded with many thousands of interlocking steel containers are bringing electronic and kitchen gadgets, socks and pants, garden tools and sports equipment from Asia to the shopping centers of Europe and North America, as well as to hawkers of cheap clothes and kitchenware in Africa and Latin America. Giant tankers move crude oil from Saudi Arabia to refineries in India and Japan, and liquefied natural gas from Texas to storage tanks in France and South Korea. Large bulk carriers full of iron ore leave Brazil for China and return empty (as do the oil tankers) to their home ports. Apple's American-designed iPhones are assembled in a Taiwanese-owned factory (Hon Hai Precision, trading as Foxconn) in Shenzhen, in China's Guangdong province, from parts coming from more than a dozen countries, and the phones are then distributed globally in a highly choreographed feat of integrated engineering and marketing.[1]

International migrations include families from Punjab or Lebanon arriving in Toronto and Sydney on scheduled jet flights; migrants risking their lives in rubber dinghies as they try to reach Lampedusa or Malta; and young adults seeking higher education abroad in London, Paris, or in small colleges in Iowa and Kansas.[2] Travel for leisure has reached such levels that in many cases the pre-pandemic label of "overtourism" was but a mild description of what was taking place in Rome's San Pietro, where the basilica was jammed with selfie stick–wielding tourists on quick package European tours, or on Asian beaches that became so degraded they had to be closed to visitors.[3] The COVID-19 pandemic's onset led to new acute overtourism crises, as hundreds of elderly people were shut in on cruise ships off the shores of Japan or Madagascar in the early spring of 2020—and yet before the end of the year, even as new waves of infection were rising fast around the world, major companies advertised new megaship cruises for 2021 (such is modern restlessness!).

Statistics concerning money movements greatly underestimate the real (including massive illegal) flows. The global merchandise trade is now close to $20 trillion a year, and the annual value of world trade in commercial services is close to $6 trillion.[4] Global foreign direct investment doubled between 2000 and 2019 and now approaches $1.5 trillion a year, while by 2020 global currency trading totaled nearly $7 trillion per day.[5] And numbers describing global information flows are many orders of magnitude higher than these money transfers—not just in terabytes or petabytes but in exa (10^{18}) and yotta (10^{24}) bytes of data.[6]

Obviously, understanding how the modern world really works cannot be done without appreciating the evolution, the extent, and the consequences of this multifaceted process which entails (according to what I think is perhaps the best concise definition) "the growing interdependence of the world's economies, cultures, and populations, brought about by cross-border trade in goods and services, technology, and flows of investment, people, and information."[7] Contrary to widely held beliefs, the process is not new; moving jobs to countries with low labor costs (labor arbitrage) is just one of its several requisite drivers; and there is nothing inevitable about its future expansion and intensification. Perhaps the greatest misconception about globalization is that it is a historical inevitability preordained by economic and social evolution. Not so—globalization is not, as a former US president claimed, "the economic equivalent of a force of nature, like wind or water"; it is just another human construct, and there is now a growing consensus that, in some ways, it has already gone too far and needs to be readjusted.[8]

In this chapter I will show that globalization is a process with considerable history (although, in the past, the rising flows of goods, investment, and people were not subsumed under the label), and the recent attention to the phenomenon escalated due to its extent, not because of its novelty. Google's Ngram Viewer charts provide excellent illustrations of long-term trends of attention paid to any notable developments. The chart for globalization consists of a near-zero flat line until the mid-1980s, then a steep rise of interest during the next two decades—a 40-fold rise of frequency between 1987 and 2006, when interest peaked—followed by a 33 percent decline by 2018.

If low labor costs were the sole reason for locating new factories abroad—as many people seem to erroneously believe—then sub-Saharan Africa would be the most obvious choice, and India would almost always be preferable to China. But during the second decade of the 21st century, China averaged about $230 billion of foreign direct investment a year, compared to less than $50 billion for India and just around $40 billion for all of sub-Saharan Africa (excluding South Africa).[9] China provided a combination of other attractors—above all, centralized one-party government that could guarantee political stability and acceptable investment conditions; a large, highly homogeneous and literate population; and an enormous domestic market—that made it the preferred choice over Nigeria, Bangladesh, and even India, resulting in a remarkable collusion between the world's largest communist state and a nearly complete lineup of the world's leading capitalist enterprises.[10]

Globalization has been linked, approvingly, with the advantages, benefits, creative destruction, modernity, and progress it has brought to entire nations. China has been by far its greatest beneficiary, as the country's reintegration into the global economy helped to reduce the number of people living in extreme poverty by 94 percent between 1980 and 2015.[11] But these gains and praises coexist with various degrees of disapproval or even outright rejection of the process, with the discontent and anger that have resulted from the loss of well-paying jobs to offshoring (with post-2000 losses especially prominent in several sectors of the US economy), from the race to the bottom as labor arbitrage drives remunerations ever lower, and from growing inequality and new kinds of immiseration.[12]

While there is much with which to agree, and disagree, in these reactions and analyses, this chapter will be neither a reiteration of often-told narratives that have crowded economic publications of the past two generations, nor a polemic about the desirability of the phenomenon. My goal is to explain how technical factors—above all, new prime movers (engines, turbines, motors) and new means of communication and information (storage, transmission, and retrieval)—made successive waves of globalization possible, and then to point out how these technical advances have been contingent on the prevailing political and social conditions. As a result, there is nothing inevitable about the continuation and further intensification of the process, and the significant, decades-long, post-1913 retreat from globalization, as well as recent reversals and concerns about the security of existing supply chains, serve as obvious reminders of this reality.

Globalization's distant origins

In its most fundamental physical way, globalization is, and will remain, simply the movement of mass—of raw materials, foodstuffs, finished products, and people—and the transmission of information (warnings, guidance, news, data, ideas) and investment within and among the continents, enabled by techniques that make such transfers possible on large scales and in affordable and reliable ways. Inevitably, these transfers entail energy conversions, and while moving mass and transmitting information can be done by deploying human and animal muscles (carrying loads, sending messengers on horseback), these animate prime movers have very limited power, endurance, and range—and, of course, they are incapable of bridging the oceans.

Sails, going back to the Egypt of more than 5,000 years ago, were the first inanimate energy converters to make such links possible, but only steam engines, aided by better means of navigation, brought the large-scale, low-cost, and reliable interchange—and only with the post-1900 diffusion of internal combustion engines (on land, on the ocean, and in the air) and the post-1955 adoption of solid-state (semiconductor) electronics has this process increased to unprecedented levels. But these innovations intensified globalization; they didn't launch it. The process (unlike its post-1985 prominence) is not a new phenomenon, and in this chapter I will trace both the timing and the extent of its past waves—and the limits of their eventual reach and intensity.

The process began a long time ago, but its first rounds were inherently limited. Trading obsidian along prehistoric routes in parts of the Old World more than 6,000 years ago is not, as has recently been claimed, an example of globalization[13]—but many ties before the European "discovery" of America were relatively intensive and truly intercontinental. Ships sailed regularly from Berenike, the Red Sea port in Roman Egypt, to India, as they did from Basra: Cassius Dio wrote in 116 CE how the emperor Trajan, during his temporary occupation of Mesopotamia, stood at the Persian Gulf's shore watching a ship leaving for India and wishing that he were as young as Alexander, who had led his armies to that faraway country.[14] Chinese silk made it, via the Parthian Empire, all the way to Rome—as did regular shipments of grain and extraordinarily heavy loads of ancient obelisks from Egypt, and wild animals from Mauretania Tingitana (northern Morocco).[15]

But the scattered linking of parts of Europe, Asia, and Africa is a far cry from a truly global reach. Only the inclusion of the New World (starting in 1492) and the first circumnavigation of the Earth (1519) began to satisfy this definition, and a mere century later commercial exchanges tied European states with the interior of Asia, India, and the Far East, as well as with coastal regions of Africa and with both Americas—and only Australia was still left aside. Some of these early links were as enduring as they were transformative. The East India Company, headquartered in London and operating between 1600 and 1874, traded a wide range of items—largely to and from the Indian subcontinent—ranging from textiles and metals to spices and opium. The Vereenigde Oost-Indische Compagnie (Dutch East India Company) imported spices, textiles, gems, and coffee mostly from Southeast Asia; it kept its uninterrupted monopoly on trade with Japan for two centuries (between 1641 and 1858), and the Dutch domination of the East Indies ended only in 1945.[16]

At the same time, technical capabilities put clear limits on the frequency and intensity of these early exchanges, and here I will use their key markers—the maximum power and speed of individual modes of transport, and the ability to communicate over long distances ever more quickly and reliably—to trace the four distinct eras of globalization.

Incipient globalization eventually connected the world with far-flung but not very intensive exchanges enabled by sail ships. Steam engines made these linkages more common, more intensive, and much more predictable, while telegraph provided the first truly global means of (near-instant) communication. The combination of the first diesel engines, flight, and radio elevated and accelerated these enablers of globalization. And large diesels (in shipping), turbines (in flight), containers (enabling intermodal transport), and microchips (allowing unprecedented controls thanks to the volume and speed of information-handling) brought globalization to its highest stage.

Wind-driven globalization

Starting at the beginning, the limits of globalization dependent solely on animate power are easily stated. Human and animal muscles were the only prime movers on land, restricting the weight of goods that could be carried by porters (maxima of 40–50 kilograms) or by animal caravans (horses or camels, loads of 100–150 kilograms per animal) and limiting their daily progress.[17] Caravans on the Silk Road (from Tanais on the Black Sea via Sarai to Beijing) took a year, implying an average speed of about 25 kilometers per day. Wooden sail ships undertaking long-distance voyages were far from numerous, and they had small capacities, traveled slowly, lacked accurate means of navigation, and frequently failed to complete their journeys.

Detailed records of Dutch shipping to Asia document these limits.[18] They show that the average duration of a voyage to Batavia (present-day Jakarta) was 238 days (eight months) during the 17th century, and another month from Batavia to Dejima, the small Dutch outpost in the Nagasaki harbor. And average speeds during the 18th century were slightly slower, with journeys taking 245 days. Given the distance of 15,000 nautical miles (27,780 kilometers) between Amsterdam and Batavia, this implies an average speed of 4.7 kilometers per hour, the equivalent of rather slow walking! That poor average is a result of some half-decent speeds when running before the wind (the wind coming from directly behind the vessel), and other days when ships were becalmed by the equatorial doldrums or when long spells of strong prevailing winds required laborious tacking—or giving up and waiting for the wind to shift.

During the 17th and 18th centuries the Dutch built only 1,450 new ships for Asian trade (averaging seven a year) with capacities of just 700–1,000 tons. That was good enough to make a profit carrying such high-value cargoes as spices, tea, and china, but completely uneconomical for any trade in bulk commodities (valuable Japanese copper was the major exception). And while the sailings to Batavia were limited by the availability of ships and the risks of travel, sailings to Japan were restricted by the Tokugawa shoguns to no more than 2–7 ships a year and to just one arrival a year during the 1790s. And because the Dutch East India Company kept detailed records, we also know the number of people who boarded the more than 4,700 ships bound from the Netherlands to the East Indies: nearly 1 million people made this journey between 1595 and 1795, but that is just 5,000 a year, and about 15 percent of them died before reaching Ceylon or Batavia.[19]

Even so, during the second century of the early modern era (1500–1800) the societies at the forefront of this still-modest but rising wave of globalization were influenced by these long-range interchanges.[20] Not surprisingly, given their newly acquired riches and contact with other continents, the lives of urban elites during the Dutch Republic's Golden Age (1608–1672) offer perhaps the best examples of these new benefits. Their growing range of possessions and experiences were the obvious markers of the gains derived from trade and from material and cultural exchanges, and many famous painters provide a fascinating record of this incipient affluence.

Works by Dirck Hals, Gerard ter Borch, Frans van Mieris, Jan Vermeer van Delft, and many lesser masters show these new profits turned into tiled floors, glass windows, well-made furniture, thick tablecloths, and musical instruments.[21] Some have argued that this can all be dismissed because this genre of painting depicted a fantasy world that never existed in reality.[22] Exaggeration and stylization were certainly present, but, as historian Jan de Vries makes clear, what he calls the "New Luxury" (generated by urban society) was real: not striving for grandeur and excess, but evinced in products of good craftsmanship—from furniture to tapestries, from Delft tiles to silver utensils—including about 3 million paintings owned by families in Holland in the 1660s.[23]

And there were other, more direct, proofs of reaching out and bringing in: the presence of Africans in Amsterdam, the popularity of maps, the profitable business of compiling and publishing atlases, the consumption of sugar and exotic fruits, the import of spices (the Dutch colonization of the East Indies began in 1607 by taking over Ternate, the largest producer of cloves, followed shortly afterwards by occupying the nutmeg-growing Banda Islands), and the drinking of tea and coffee.[24]

But these early exchanges had limited economic impact, as they never reached beyond the small segments of people who benefited from the new ventures. The countryside was left with its traditional ways. This was just an incipient, selective, and limited globalization without any substantial nationwide impacts, to say nothing about truly global consequences. For example, economist Angus Maddison estimated that in 1698–1700 commodity exports from the East Indies accounted for just 1.8 percent of the Dutch net domestic product, and that the Indonesian export surplus was a mere 1.1 percent of the Dutch GDP—and nearly a century later (1778–1780) both of these shares were still only 1.7 percent.[25]

Steam engines and telegraph

The first quantitative leap in the process of globalization came only with the combination of more reliable navigation, steam power (resulting in larger ship capacities and faster speeds), and the telegraph—the first means of (nearly) instant long-distance communication. Navigation came first, in 1765, with John Harrison's fourth highly accurate sea clock, a chronometer which made it possible to determine exact longitude. But the leap in speeds and capacities had to wait until steam engines displaced sails in intercontinental shipping, when screws made paddlewheels obsolete, and when steel-hulled ships became dominant.[26]

The first steam-powered westward transatlantic crossings took place in 1838, but sailing ships remained competitive for another four decades. With wind as the prime mover, the cost of carrying a unit of cargo per unit of distance by a sailing ship was largely independent of the length of the voyage; while the longer the steamship voyage, the more of the vessel's deadweight capacity had to be loaded with coal to fuel relatively inefficient engines, leaving less room for cargo. Refueling stations reduced, but did not eliminate, this disadvantage.[27]

This long coexistence of sail and steam is well documented by the German transition: by 1873, sailing ships had lost the competition on the intra-European routes, while on the intercontinental routes sails had the advantage until 1880 but lost it rapidly afterwards with the adoption of more efficient engines.[28]

All pioneering steamships crossing the Atlantic were propelled by paddlewheels, but screw propulsion was commercially introduced during the 1840s; and in 1877 Lloyd's Register of Shipping approved steel as an insurable construction material just as new production methods made the metal abundant and affordable (see chapter 3). Steel hulls and screws and large steam engines made it possible to cover reliably 30 and then 40 km/hour compared to the average of 20 km/hour for the fastest clippers of the 1850s, and long-distance shipping also gained new markets with the exports of live cattle and—starting in the 1870s—chilled meat (carried almost exclusively by passenger liners) and butter from the US, Australia, and New Zealand.[29]

Practical telegraph was developed during the late 1830s and the early 1840s; the first (short-lived) transatlantic link cable was laid in 1858; and by the century's end undersea cables had connected all continents.[30] For the first time in history, trading could take into consideration the knowledge of demand and prices in different parts of the world—and the availability of a new powerful prime mover could translate this information into profitable international exchanges: when the price of Iowa beef was cheaper than British beef of inferior quality and new refrigerating techniques became available, for example, the exports of frozen American meat rose rapidly—more than quadrupling between the late 1870s and the late 1900s.

During this wave of steam-driven globalization, the role of the telephone—a device far superior to telegraph for direct personal communication—remained limited.[31] Its patenting and the first public demonstration in 1876 were followed by a slow diffusion of service mediated by manual exchanges. US telephone ownership rose from less than 50,000 in 1880 to 1.35 million by 1900 (one phone for every 56 Americans); calling distances expanded gradually (a call from New York to Chicago could be made only in 1892); the first transcontinental calls to San Francisco (via multiple exchanges) came in 1915; and a three-minute conversation cost about $20, or more than $500 in 2020 monies. The first intercontinental call—from the US to the UK—came only in 1927, and even the monopolized domestic service remained relatively expensive for the next two generations.[32]

But the advances in intercontinental shipping, combined with rapid post-1840 construction of railroads—across Europe and North America, as well as in India, other regions of Asia, and Latin America—created the first wave of a truly large-scale globalization. The total volume of global trade quadrupled between 1870 and 1913; the share of trade (exports and imports) in the worldwide economic product rose from about 5 percent in 1850 to 9 percent by 1870, and to 14 percent in 1913; and the best estimates for 13 countries (including Australia, Canada, France, Japan, Mexico, and the UK) show their combined share rising from 30 percent in 1870 to 50 percent just before the First World War.[33]

Large steam liners were also able to move passengers on an unprecedented scale. During the sailing era, packet ships carried 250–700 steerage passengers; by the first decade of the 20th century, a steam liner might carry more than 2,000 passengers.[34] Leisure travel, a form of temporary migration formerly reserved for the privileged classes, took off in its many manifestations with steam-powered trains and ships. With Thomas Cook leading the way in 1841, travel agencies offered package tours, and spa and seaside vacations became fashionable as people visited Baden-Baden, Karlsbad, and Vichy, and traveled to Trouville on the French Atlantic coast or to Capri.

Some of these trips were transcontinental: affluent Russian families took trains all the way from Moscow and Saint Petersburg to the French Riviera. Some travelers sought physical challenges (newly fashionable Alpine mountaineering), while others went on (more affordable) religious pilgrimages.[35] And this new mobility also had a distinctly political dimension, with exiles—traveling by trains and ships—seeking refuge in foreign countries: most famously, nearly all future prominent Bolshevik leaders (Lenin, Leon Trotsky, Nikolai Bukharin, Grigory Zinoviev) spent many years abroad in Europe and the US.[36]

And I think it is quite reasonable to argue that steam globalization also helped to create a new kind of literary sensibility, with Joseph Conrad (Józef Korzeniowski) as its master. The protagonists of his three greatest novels find themselves far from their homes thanks to the era's mass trade and travel (Nostromo in South America, Jim in Asia, Marlow in Africa), and their lives and misfortunes were linked to steamships: Nostromo, in the eponymous novella, is known as Capataz de Cargadores (Head Longshoreman); Jim's life takes a tragic turn while he is helping transport Muslim pilgrims from Asia to Mecca in *Lord Jim*; and Marlow's transformation could not have taken place without Western goods being brought deep into the Congo basin in *Heart of Darkness*.

The first diesel engines, flight, and radio

The next fundamental advance in prime movers that raised the capability of long-distance shipping was the replacement of steam engines with diesel engines—machines of superior efficiency and reliable performance.[37] Two concurrent processes that promoted further globalization were the invention of airplanes powered by reciprocating gasoline engines, and radio communication. The first brief flights—by the Wright brothers—took place in late 1903; hundreds of planes flew in combat during the First World War; and the first airline company, the Dutch KLM, was set up in 1921.[38] The first transatlantic radio signal arrived in December 1901; the French army deployed the first portable transmitters for air-to-ground communication in 1916; and the first commercial radio stations began to broadcast in the early 1920s.[39]

Rudolf Diesel deliberately set out to design a new, more efficient, prime mover, and by 1897 his first (heavy and stationary) engine had reached an efficiency of 30 percent, double the performance of the best steam engines.[40] But the first marine engine was installed only in 1912 on *Christian X*, a Danish freighter. Diesel-powered ships carried much less fuel than coal-fired steamers, but could travel further without refueling because the new engines were nearly twice as efficient—and because, per unit of mass, diesel oil contains nearly twice as much energy. An American engineer who saw the first diesel-powered vessel after her maiden voyage to New York in 1912 concluded that: "marine history is being written by the advent of the Diesel engine."[41]

By the 1930s, when diesel engines conquered the shipping market, the rapidly maturing aviation industry began to deliver the first airplanes capable of flying profitably over long distances. In 1936 came the first deliveries of the Douglas DC-3, a twin-engine aircraft capable of carrying up to 32 passengers just a bit faster than the landing speed of modern jetliners.[42] Three years later came the Boeing 314 Clipper, a long-range flying boat with an impressive range of 5,633 kilometers—still not enough to cross the Pacific, but more than enough to reach Honolulu from San Francisco before continuing to Midway, Wake, Guam, and Manila to reach Asia.

The Clipper had no shortage of physical comforts for its 74 passengers—including a stateroom and a dining room, dressing rooms, and seats that converted into bunks—but there was no way to eliminate the noise and vibration of reciprocating engines, and the highest cruising altitude (5.9 kilometers) was still too low to put it above the most turbulent atmospheric layers. With three stops it took 15½ hours from New York to Los Angeles, and the first London to Singapore link in 1934 took eight days with 22 layovers, including Athens, Cairo, Baghdad, Basra, Sharjah, Jodhpur, Calcutta, and Rangoon.[43] But long as it was, it was a considerable improvement on the approximately 30 days needed to travel by ship from Southampton via the Suez Canal.

Radio was of critical importance for better sea and air navigation, and, compared to the telegraph, it was also a superior tool for mass-scale dissemination of instant information. Radio communication was deployed first on transatlantic ocean liners: thanks to *Titanic*'s distress message—"CQD Titanic 41.44 N 50.24 W," sent at 12:15 a.m. on April 15, 1912—700 people in lifeboats were saved by *Carpathia*.[44] Radio navigation made great advances during the 1930s with the introduction of range stations: planes on course to an airport heard a continuous audio tone; those drifting off-course heard N Morse code (– •) when to the left of the path, A (• –) when right of it.[45]

Wireless broadcasts required no expensive undersea cables, and they could achieve wide area coverage and universal access (anybody with a simple receiver could listen). Not surprisingly, the adoption of radio receivers was rapid: within a decade after their introduction, 60 percent of American families had them—nearly as fast a rate of acquisition as that of black-and-white TVs (which also originated in the 1920s) after the Second World War, and a faster rate than the subsequent diffusion of color TV which in the US took off rapidly during the early 1960s.[46]

Marine diesels and reciprocating aircraft engines remained the technical enablers of globalization during the two interwar decades, and their mass-scale deployment made the decisive contribution to the outcome of the Second World War. By the conflict's end, the US had built nearly 296,000 airplanes compared to about 112,000 in Germany and 68,000 in Japan.[47] In 1945, the United States emerged as the world's dominant power, and the economic recovery of Western Europe was fast. Aided by the US investment (1948's Marshall Plan), all of the region's countries surpassed their prewar (1934–1938) level of industrial production by 1949, while Japan's recovery was accelerated by the contribution of that country's industries to the Korean War.[48]

The stage was thus set for a period of unprecedented growth and integration, as well as for extensive social and cultural interactions. Communist economies, led by the USSR and China, were the notable exceptions: although they reported impressive economic growth rates, they were highly autarkic and operated with very little foreign trade beyond their bloc (and kept their citizens from traveling abroad).

Large diesels, turbines, containers, and microchips

This distinct and intensive, but still far from universal, spell of post-1950 globalization—which ended in 1973–1974 with OPEC's two rounds of oil price increases and which was followed by 15 years of relative stagnation—was enabled by a combination of four fundamental technical advances. These were the rapid adoption of much more powerful and efficient designs of diesel engines; the introduction (and even faster diffusion) of a new prime mover, the gas turbine used for the propulsion of jetliners; superior designs for intercontinental shipping (massive bulk carriers for liquids and solids, and the containerization of other cargoes); and quantum leaps in computing and information processing.

These advances took off with the first electronic computers—which used unreliable and bulky vacuum tubes and were built during, and right after, the Second World War—and their progress was revolutionized by the patenting (1947–1949) and commercialization (starting in 1954) of the first transistors, devices that remain the foundation of modern solid-state electronics. The next step (late 1950s–early 1960s) was placing increasing numbers of transistors on a microchip to create integrated circuits, and in 1971 Intel released its 4004, the world's first microprocessor. It contained 2,300 transistors, forming a complete general-purpose central processing unit suitable for many programmable applications.

And despite the recent perceptions of the transformative nature of technical capabilities deployed since the beginning of the 21st century (above all, advances in artificial intelligence and synthetic biology), our world is still beholden to these critical pre-1973 achievements. Moreover, as there are no immediately available alternatives that could be deployed for the same tasks on similarly massive scales, we will depend on these techniques—be they giant marine diesel engines, container ships and wide-body jetliners, or microprocessors—for decades to come. And that is also why these techniques deserve a closer look.

The scale of global economic expansion between 1950 and 1973 is best illustrated by the growing output of the four material pillars of modern civilization (for their assessment, see chapter 3) and by the world's rising energy demand (see chapter 1).[49] Steel production nearly quadrupled (from about 190 to 698 megatons per year), cement production increased almost sixfold (from 133 to 770 megatons), ammonia synthesis almost eightfold (from less than 5 to 37 megatons of nitrogen), and plastic output was more than 26 times higher (from less than 2 to 45 megatons). Primary energy output nearly tripled, and crude oil consumption rose almost sixfold as the world became increasingly dependent on Middle Eastern oil. As a result, there is no contest as to which technique has made the greatest difference in enabling mass-scale transport in the global economy: without diesel engines, intercontinental trade in bulk cargoes—from grain to crude oil—would have been just a small fraction of recent shipments.

After the Second World War, crude oil tankers were the first vessels to grow in capacity as the rapid economic growth of Western Europe and Japan coincided with the availability of newly discovered Middle Eastern giant oil fields (Saudi Arabia's Ghawar, the world's largest, was found in 1948 and began flowing in 1951), and exports of this inexpensive fuel (until 1971 it sold for less than $2 per barrel) required vessels of increasing capacities. Typical pre-1950 oil tankers had capacity of only 16,000 deadweight tons (mostly the ship's cargo, but also its fuel, ballast, provision, and crew). The first tanker of more than 50,000 deadweight tons was launched in 1956, and by the mid-1960s Japanese shipyards began launching very large crude carriers (VLCC) with capacities of between 180,000 and 320,000 deadweight tons. Beyond that came ultra large crude carriers (ULCC) and seven ships larger than 500,000 deadweight tons were launched during the 1970s, too large to allow for flexible routing as they can be accommodated only at the deepest ports.[50] This growing fleet made it possible to increase the shipments of Middle Eastern oil from less than 50 megatons in 1950 to about 850 megatons by 1972.[51]

Even as crude oil exports were soaring during the late 1950s and early 1960s, there was no way to ship natural gas, a cleaner fuel than coal or refined oil fuels and also well suited both to industrial and household uses and to the highly efficient generation of electricity.

Intercontinental natural gas shipments became possible with the introduction of the first liquefied natural gas (LNG) tankers (carrying the fuel at −162°C in insulated containers), which brought exports from Algeria to the UK starting in 1964 and from Alaska to Japan in 1969.[52] But for decades, the ships were of low capacity and the market was limited to long-term contracts with a small number of buyers.

Growing intercontinental trade required new modes of specialized shipping. Bulk carriers with large compartments and massive watertight hatches were designed to transport coal, grain, ores, cement, and fertilizer, and they could be rapidly loaded and unloaded. But the greatest shipping innovation came in 1957, when a North Carolina trucker Malcolm McLean finally transformed his pre–Second World War idea—carrying cargo in uniformly-sized steel boxes, which are easy to load by large port cranes and can be off-loaded directly onto waiting trucks or trains or stacked temporarily for later distribution—into a commercial reality.

In October 1957, *Gateway City*, a freighter whose hold was fitted with cellular compartments to accommodate 226 stacked containers, became the world's first true container ship, and McLean's Sea-Land company began a regular container service to Europe (Newark–Rotterdam) in April 1966 and to Japan in 1968.[53] New ships were also needed for expanding intercontinental car exports. The American market opened up first to Volkswagen's Beetle (the first car imported already in 1949) and then to small Japanese designs (the Toyopet since 1958, Honda N600 since 1969, and Honda Civic since 1973), and new roll-on/roll-off vessels (mostly with built-in retractable loading ramps) were designed to meet those needs. After years of slow adoption, VW's sales peaked at 570,000 units in 1970, and Japanese designs continued to gain US market share over the coming decades.[54]

Fortunately, there was no problem meeting the propulsion needs of these new large ships. The pre–Second World War sizes of the largest diesel engine had more than doubled by the late 1950s—to more than 10 megawatts—as their efficiencies approached 50 percent.[55] Subsequently, the maximum power of these massive multi-cylinder engines rose to 35 megawatts by the late 1960s and to more than 40 megawatts by 1973. Any diesel engines rated above 30 megawatts can power the largest ULCC, and hence the size of these vessels has never been limited by the availability of adequate prime movers.

The quest for a practical gas turbine, a radically new prime mover in which fuel is sprayed into a stream of compressed air in order to generate a high-temperature gas that expands and exits the machine at high speed, resulted in the first stationary turbine (for electricity generation) in 1938, just as the first practical designs for jet engines emerged—independently and at nearly the same time—in prewar England and Germany.[56] Frank Whittle and Hans von Ohain were the first engineers to test turbines that were sufficiently efficient and reliable to power military airplanes. Small numbers of these jets were used in combat in late 1944, too late to have any effect on the predetermined course of the war, but after its end the British industry pressed its advantage and in 1949 the Comet became the world's first commercial jetliner, powered by four de Havilland Ghost turbojet engines.[57]

Unfortunately, by 1954 a series of fatal accidents (not related to the engines) forced the withdrawal of the plane from service, and when a redesigned Comet returned in 1958 it was quickly overshadowed by Boeing's 707, the first design of a still-expanding family of jetliners.[58] The second in line was the tri-engine Boeing 727, and in 1967 came the Boeing 737, the smallest plane of the series. In 1966 William Allen, the company's chairman, took a bold decision—investing more than twice the company's worth, and hence essentially betting its future on the success of the project—to develop the first wide-body jetliner.

Supersonic jets were expected to take over intercontinental routes—the development of the British-French Concorde began in 1964—but supersonic flight remained restricted to the expensive and noisy Concorde and it was Boeing's 747 that became the most revolutionary plane design in history.[59] The plane was actually conceived as a freighter: its wide body allowed the placing of two standard ship containers side by side, and the cockpit in the top bubble made it possible to turn up the nose for front-loading. The prototype took off less than three years after Pan Am's order for 25 of the 747s, and the first commercial flight left New York for London on January 21, 1970.

The plane's size (a maximum takeoff weight of 333 tons) was made possible by deploying four Pratt & Whitney turbofan engines.[60] Unlike turbojet engines, where all compressed air moves through the combustion chamber, in turbofans larger masses of less compressed and hence slower-moving air bypass the combustor and help to generate higher thrust during takeoff (and do so with less noise). Engines on 707s had a bypass ratio of 1:1, on the 747 it was 4.8:1, with nearly five times as much air bypassing the turbine.

Total deliveries of the 747 have added up to 1,548 planes in half a century of production, and Boeing estimates that during those five decades the planes carried 5.9 billion people, the equivalent of about 75 percent of the world's population.[61] The plane's revolutionary design changed intercontinental travel, as wide-body jets have transported hundreds of millions of people to a growing number of destinations with steadily declining costs and increasing safety.

The integration of the global economy has been closely tied to the introduction of wide-body jetliners—to the Boeing 747 and to its later Airbus (A340 and A380) emulators. Their services have been particularly important for Asian exporters, who use them to deliver on short notice many highly sought-after or seasonal items (the latest mobile phone brands, Christmas gifts) to North American and European markets. And wide-body airplanes enabled mass-scale tourism to previously rarely visited destinations (runways long enough to accommodate 747s are in Bali and Tenerife, Nairobi and Tahiti), intercontinental immigration trips, and educational exchanges.

Of course, the advances of globalization were closely tied not only to the rising capacities and better performance of powerful prime movers but also to the relentless miniaturization of components needed for computing, information processing, and communication. The development of radio, and later of television and the first electronic computers, depended on deploying a variety of vacuum tubes, beginning with diodes and triodes during the first decade of the 20th century. Four decades later our reliance on these large assemblies of hot glass became a limiting factor for the development of electronic computing.

ENIAC, the first electronic general-purpose digital computer, had 17,648 vacuum tubes, a volume of about 80 cubic meters (the footprint of roughly two badminton courts), with its power supply and cooling system it weighed about 30 tons, and its frequent operating interruptions were caused by recurrent failures of tubes that required near-constant maintenance and replacement.[62] The first practical transistors—solid-state devices performing the same functions as glass-enclosed devices—became commercially available during the early 1950s, and before the decade's end the ideas of several American inventors (Robert Noyce, Jack Kilby, Jean Hoerni, Kurt Lehovec, and Mohamed Atalla) resulted in the production of the first integrated circuits, with active (transistors) and passive (capacitors, resistors) components built and interconnected on a thin layer of silicon (a semiconducting material). These circuits can perform any specified computing functions, and their first practical uses were in rockets and space exploration.[63]

The next critical step was taken by Intel in 1969 when it began to design the world's first microprocessor, emplacing more than 2,000 transistors on a single silicon wafer in order to perform a complete set of prescribed functions: in the case of Intel's pioneering 4044, it was to run a small Japanese electronic calculator.[64] The 4044 founded Intel's decades-long prominence in microchip design that led to the first personal computers (the relatively expensive, slow and heavy desktops of the late 1970s and early 1980s) and portable electronics ranging from mobile phones (the first costly designs of the late 1980s) to laptops, tablets, and smartphones.

The years between 1950 and 1973 were marked by rapid economic growth in virtually every part of the world: its global annual mean rate and its average per capita gains were nearly 2.5 times greater than during the previous globalization wave of 1850–1913, and the value of exported goods in the world economic product rose from a low of just over 4 percent in 1945 to 9.6 percent in 1950 and about 14 percent in 1974, equaling the 1913 share but with trade volume nearly ten times higher.[65] Economic growth was nearly universal (China's Great Famine years of 1958–1961 were the most consequential exception) but the benefits of this golden age of economic expansion—the postwar rebound, with high rates of growth helping to decrease economic inequality—were disproportionately concentrated in the West: by

1973, North America and the countries of Western Europe accounted for more than 60 percent of global exports.[66] As the major Western European economies (Germany, UK, France) and Japan became the era's most dynamic traders, inevitably America's share in world trade gradually eroded.

While trade was expanding and consumers in Western countries enjoyed greater access to a wider variety of imports, international travel—be it for business or leisure—remained relatively limited, as did international migration and the numbers of people studying or temporarily working abroad. Germans did not fly to Thailand or Hawaii; they drove to Italian beaches. The share of immigrants in the US population, which peaked at nearly 15 percent just before the First World War, reached a new low of less than 5 percent in 1970.[67] And the suggestion that China, cut off from the world by Mao-inspired convulsions, would send masses of students to American universities would have been seen as nothing but an improbable fiction.

And then (for reasons explained in the first chapter, where I traced modern civilization's dependence on crude oil) it seemed that the postwar spell of limited but intensive globalization was over. OPEC-driven oil price increases caused globalization to falter, weaken, and recede, but this retreat did not affect all economic sectors—and in a matter of years a combination of effective adjustments laid foundations for a new round of globalization that, thanks to new political alignments, progressed further than any of the preceding waves.

Enter China, Russia, India

This time the expansion—enabled, as always, by technical factors—went so far because, for the first time in modern history, it could go that far. By the late 1960s, technical capabilities were ready for unprecedented global integration: energy supply was plentiful, there was no shortage of money to invest, and all that was needed was to extend the globalization process to the nations that did not participate in the first postwar round. This began, finally, when the technical and financial means became decisively augmented and potentiated by fundamental political reversals, as China, Russia, and India became major participants in global trade, finances, travel, and talent flows.

The gradual opening of China began in 1972 with Richard Nixon's visit to Beijing, took a decisive turn in late 1978 (two years after Mao Zedong's death) with the rise of Deng Xiaoping and the launching of long-overdue economic reforms (the de facto privatization of farming, modernization of industry, and partial return to private enterprise), and accelerated after China joined the World Trade Organization (WTO) in 2001. In 1972, China had no trade with the US; 1984 was the last year the US had a surplus trading goods with Beijing; in 2009, China became the world's largest exporter of goods; and by 2018 its exports accounted for more than 12 percent of all global sales, and its trade surplus with the US reached nearly $420 billion before declining by about 18 percent in 2019 due to rising tensions between the two economic superpowers.[68] But it is too early to forecast any long-term retreat in trade or return to ever-tighter economic integration.

After decades of the Cold War, the USSR began to unravel during the late 1980s. Its satellite states detached themselves first (the Berlin Wall fell on November 9, 1989), and the Soviet state was officially dissolved on December 26, 1991.[69] For the first time in history, it became possible for every major economy to become open (still to varying, but in nearly all cases unprecedented, degrees) to foreign investment, intensifying international trade, and populations previously forbidden to travel freely abroad joined mass-scale tourism and took advantage of new opportunities to emigrate and to temporarily work and study abroad. Trade expansion took place within a globally agreed framework provided by the WTO.[70]

India, with its messy electoral and multiethnic politics, has not been able to replicate China's post-1990 rise driven by the unchallenged rule of a single party, but the record of its per capita GDP growth during the first two decades of the 21st century indicates a clear departure from the previous decades of poor performance. Between 1970 and 1990, the country's per capita GDP (in constant monies) had actually declined in six separate years and stayed below 4 percent for four years, while between 2000 and 2019 there was an annual growth above 4 percent for 18 of those years.[71] Moreover, since 2008 the country's annual growth of merchandise exports has been, at 5.3 percent, only slightly behind China's 5.7 percent, and the impact of India's software engineers in Silicon Valley (where they have been the single most important contingent of skilled immigrants in the industry) has been far above Chinese contributions.[72]

India's rise has coincided with the marginalization of the Congress party that ruled the country for decades after it gained its independence in 1947, while both Russia and China have retained many attributes of central economic and social control. Unlike in the new, nationalist, Russia, the Communist Party remains in firm command in China, but both countries have allowed (with notable repressive exceptions) freedom of travel that has led to new waves of tourists—with favorite destinations being the Mediterranean countries for the Russians; Thailand, Japan, and Europe for the Chinese—and there has been an unprecedented influx of Chinese, Indian, and South Korean students to the West, above all to the US.

International trade's share in the world economic product rose from about 30 percent in 1973 to nearly 61 percent in 2008, while the total volume of trade (in constant monies) increased almost exactly sixfold, with most of the rise taking place since 1999.[73] The financial crisis of 2008–2009 cut the total volume by a tenth and trade's share of economic output by about 15 percent in 2009, but by 2018 overall trade was 35 percent above the 2008 peak and trade's share of the world economic product was back above 59 percent—and the numbers changed little in 2019. Foreign direct investment (measured as net outflows per year) is another obvious marker of globalization. In 1973 its global total was less than $30 billion (about 0.7 percent of the global economic product); two decades later it rose to $256 billion; but by 2007 it soared to $3.12 trillion (nearly 5.5 percent of the global product), a 12-fold increase in just 14 years, with Asia (and above all China) as the main destination.[74]

A Russian team measured the progress of post-2000 globalization by combining all key markers—that is by analyzing changes in the trade in goods, trade in services, and accumulated stocks of bilateral foreign direct investment (particularly important for China), and also migrants (nonexistent in China, but important for the American economy).[75] Not surprisingly, the results show the largest gains for previously isolated Russia, other former European communist economies, and China, and also for India, some African countries, and Brazil. Moreover, as a result of these shifts, by 2017 the global connectedness of China was as high as that of Japan; Russia's rivaled that of Sweden; and India could be compared with Singapore. If any of these pairings strikes you as questionable, just think of China's place as the largest manufacturer of consumer goods, of Russia's enormous exports of energy and minerals, and of the (already noted) contingent of Indian software engineers in Silicon Valley.

Globalization's multiples

Perhaps the best way to appreciate the technical advances that made this truly unprecedented globalization possible is to express their progress as multiples of their capacities, ratings, efficiencies, or performances. As already explained, the technical foundations for this dizzying round of globalization were laid before 1973, but its extent and intensity since then has demanded enormous investment in prime movers (combustion engines and electric motors in transportation) and in essential infrastructures (ports, airports, containerized shipping). As a result, we have not only more of them, but their average capacities (power, volume, throughput) have become larger, while their typical efficiencies and reliabilities have become better. So let's look at the advances that have taken place in shipping, flying, navigating, computing, and communicating since the early 1970s.

Post-1973 globalization has more than tripled the mass of seaborne trade and brought major shifts in its composition.[76] While in 1973 tanker traffic (dominated by crude oil and refined products) accounted for more than half of the transported total, in 2018 goods amounted to about 70 percent, a shift reflecting not only the rise of Asia—and, above all, China—as the world's leading source of consumer goods, but the overall increase in integration and interdependence: German carmakers assemble vehicles in Alabama, Texas-made chemicals (taking advantage of the boom in the extraction of natural gas) provide feedstocks for EU industries, Chilean fruits are exported to four continents, and Somali camels are shipped to Saudi Arabia.

This tripling of shipped mass between 1973 and 2019 required (when measured in deadweight tons) a near quadrupling of the global merchant fleet capacity. Deadweight tonnage of oil tankers slightly more than tripled, the tonnage of container ships increased about 4.5 times, and the size of the global container fleet expanded roughly 10-fold in 45 years to 5,152 ships in 2019. This order-of-magnitude increase was accompanied by a massive shift of container activity to China: in 1975, China had no container traffic and US and Japanese ports accounted for nearly half of the global activity; in 2018, China (including Hong Kong) held a 32 percent share, while the combined US and Japanese share was less than 10 percent.

As for maximum ship sizes, in 1972 and 1973 Malcolm McLean launched his largest container vessels, each with a capacity of 1,968 standard steel containers (nearly five times larger than his first converted vessels in 1957). In 1996, *Regina Maersk* could load 6,000 standard units; by 2008 the maximum was 13,800; and in 2019 the Mediterranean Shipping Company put into service six giant vessels each able to carry 23,756 standard containers, hence a 12-fold increase of maximum vessel capacity between 1973 and 2019.[77] Inevitably, this mass-scale conversion to container shipping required the commensurate conversion of freight trains and truck transportation, and these intermodal chains now bring the goods from a city in China's interior to the loading dock of a Walmart in Missouri.

And when the shipment of expensive food or flowers (freshly caught tuna from Atlantic Canada to Tokyo; green beans from Kenya to London; roses from Ecuador to New York) or high-value electronics require speed, they go by air. The belly hold of every passenger airplane carries goods, as does the growing fleet of air freighters: as a result, between 1973 and 2018 global airfreight (expressed in ton-kilometers) rose about 12-fold, while scheduled passenger traffic rose from about 0.5 trillion to more than 8.3 trillion passenger-kilometers, nearly a 17-fold gain.[78] Nearly two-thirds (5.3 trillion passenger-kilometers) of the latest total were on international flights—the equivalent of flying nearly half a billion people a year from New York to London and back again.

An increasing share of these flights is being taken by international tourists. In the early 1970s their annual global total (dominated by Americans and Western Europeans) was below 200 million; by 2018 the new record reached 1.4 billion.[79] Europe remains the main tourist destination, accounting for half of the total arrivals—with France, Spain, and Italy being the continent's most visited countries. For generations the US led overall tourist expenditures, but it was surpassed by China in 2012 and five years later Chinese tourists were spending twice as much as Americans. The rather sudden multiplication of arrivals and their disproportionate concentration in several major cities (Paris, Venice, Barcelona) has led to complaints by their permanent residents and to the first moves to limit the numbers of daily or annual visitors.[80]

The long reach of Moore's law

Increases in moving materials, products, and people, as well as the necessity of delivering materials or components just in time for new industries working without extensive inventories, were enabled (and made more reliable) by gains in navigation, tracking, computing, and communication, and much-expanded capabilities were also needed to accommodate the new deluge of international data flows. All of these advances have one fundamental technical foundation: our ability to emplace more components on an integrated circuit whose progress—doubling roughly every two years—has been, so far, conforming to the prediction made by Gordon Moore, at that time Fairchild Semiconductor's director of research, in 1965.[81]

In 1969, Moore became Intel's co-founder, and (as already noted) in 1971 the company released its first microprocessor (microchip), with 2,300 components. Microprocessor fabrication had eventually advanced from large-scale integration (up to 100,000 components) to very large-scale integration (VLSI, up to 10 million components), and to ultra large-scale integration (ULSI, up to a billion components).[82] The 10^5

mark (100,000 transistors) was reached in 1982, and in 1996, to celebrate the machine's 50th anniversary, a group of students at the University of Pennsylvania recreated ENIAC by putting 174,569 transistors on a 7.4 mm × 5.3 mm silicon microchip: the original machine was more than 5 million times heavier, it required about 40,000 times more electricity, and the recreated chip was 500 times faster.[83]

And the progress continued: the 10^8 mark was surpassed in 2003, 10^9 in 2010, and by the end of 2019 AMD released its Epyc CPU with 39.5 billion transistors.[84] This means that between 1971 and 2019 microprocessor power increased by seven orders of magnitude—17.1 billion times, to be exact. These advances were more than enough to accommodate new demands for massive data transfers (from Earth observation, spy and communication satellites, and among financial centers and data storages), instant e-mail and voice calls and highly accurate navigation.

That last capability has benefited from advances in radar detection and by the setting up and subsequent expansion and improvement of global positioning systems (GPS): the first (American) system was fully operational in 1993, and three other systems (Russia's GLONASS, EU's Galileo, China's BeiDou) followed.[85] As a result, everybody with a computer or a mobile phone can now see worldwide shipping and aviation activities in real time, just by clicking on the MarineTraffic website and watching cargo vessels (green icons) converging on Shanghai and Hong Kong, lining up to pass between Bali and Lombok, or going up the English Channel; to see tankers (red) debouching from the Persian Gulf, tugs and special craft (turquoise) serving the oil and gas production rigs in the North Sea, and fishing vessels (light brown) roaming the central Pacific (and there are many more ships there and elsewhere that do not show on the screen, because, while fishing illegally, they turn off their transponders).[86]

Analogical, and no less fascinating, is the click-away opportunity to monitor all commercial flights.[87] Early morning in Europe shows a long arc of staggered flights approaching the continent after crossing the Atlantic during the night from North and South America; evenings in North America show the long streams of jetliners following the optimal flight paths to Europe; transpacific flights to

Japan converge on Narita and Haneda during late afternoons and early evenings Tokyo time. Moreover, flight tracking makes it possible to trace changing flight paths that take into account the frequently shifting position of the jet stream.[88] Less frequent flight-path adjustments are caused by the progress of major cyclones, or by ash clouds emitted by volcanic eruptions.[89]

Inevitability, setbacks, and overreach

The history of globalization reveals an undeniable long-term trend toward greater international economic integration that is manifested by intensified flows of energies, materials, people, ideas, and information, and that is enabled by improving technical capabilities. The process is not new, but only thanks to many post-1850 innovations could it have reached its recent intensity and extent. But, as some past setbacks indicate, these technical advances do not make continued progress inevitable: most notably, the first half of the 20th century saw a significant retreat from economic globalization and hence also from the accompanying international movement of people. The reasons for this retreat are obvious, as the decades were marked by an unprecedented concatenation of large-scale tragedies and reversals of national fortunes.

The list, limited to the key events, includes the end of the Qing, China's last imperial dynasty (1912); the First World War (1914–1918); the end of Czarist Russia, when the Bolsheviks took power and years of civil war followed, ending with the establishment of the USSR (1917–1921); the unraveling of the Ottoman Empire (final dissolution by 1923); the political instability in Europe of the postwar 1920s; the stock market collapse in late October 1929; the subsequent worldwide economic crises that lasted for most of the 1930s; Japan's invasion of Manchuria (1931), the true beginning of another great war; the Nazi takeover of Germany (1933); the Spanish civil war (1936–1939); the Second World War (1939–1945); the renewed civil war in China (1945–1949); the beginning of the Cold War (1947); and Mao's proclamation of the People's Republic of China (1949). The retreat of economic globalization was substantial. The share of trade in global GDP fell from about 14 percent in 1913 to about 6 percent in 1939 and then to only 4 percent in 1945.[90]

And the accelerated post-1990 pace of globalization did not depend on just having superior technical means; it would have been impossible without concurrent major political and social transformations, most notably the post-1980 return of China to international commerce, followed (between 1989 and 1991) by the dismantling of the Soviet Empire. This means that the high degree of globalization reached during the first two decades of the 21st century has not been inevitable, and that it can be weakened by future developments. To what extent (marginally or substantially) and how rapidly (fast due to major power confrontations, gradually as a generational affair) is impossible to foresee.

Much seems to be firmly set in place. A great deal of accreted globalization, especially many changes that unfolded during the past two generations, is here to stay. Too many countries now rely on food imports, and self-sufficiency in all raw materials is impossible even for the largest countries because no country possesses sufficient reserves of all minerals needed by its economy. The UK and Japan import more food than they produce, China does not have all the iron ore it needs for its blast furnaces, the US buys many rare earth metals (from lanthanum to yttrium), and India is chronically short of crude oil.[91] The inherent advantages of mass-scale manufacturing preclude companies from assembling mobile phones in every city in which they are purchased. And millions of people will still try to see iconic distant places before they die.[92] Moreover, instant reversals are not practical, and rapid disruptions could come only with high costs attached. For example, the global supply of consumer electronics would suffer enormously if Shenzhen suddenly ceased to function as the world's most important manufacturing hub of portable devices.

But history reminds us that the recent state of things is unlikely to last for generations. British and American industries were the global leaders as recently as the early 1970s. But where are Birmingham's metal-working factories or Baltimore's steel furnaces now? Where are the great cotton mills of Manchester or of South Carolina? By

1965, Detroit's big three still had 90 percent of the US car market; now they do not have even 45 percent. Until 1980, Shenzhen was a small fishing village, when it became China's first special economic zone, and now it is a megacity with more than 12 million people: what role will it play in 2050? A mass-scale, rapid retreat from the current state is impossible, but the pro-globalization sentiment has been weakening for some time.

The accelerated deindustrialization of North America, Europe, and Japan, and the shift of manufacturing to Asia in general and to China in particular, has been the leading reason for this reappraisal.[93] This manufacturing switch has brought changes ranging from risible to tragic. In the first category are such grotesque transactions as Canada, the country with per capita forest resources greater than in any other affluent nation, importing toothpicks and toilet paper from China, a country whose wood stocks amount to a small fraction of Canada's enormous boreal forest patrimony.[94] But the switch has also contributed to tragedies, such as the rising midlife mortality among America's white non-university-educated men. There can be no doubt that America's post-2000 loss of some 7 million (formerly well-paying) manufacturing jobs—with most of that loss attributable to globalization, as most of that production moved to China—has been the principal reason of these deaths of despair, largely attributable to suicide, drug overdose, and alcohol-induced liver disease.[95]

And we now have solid quantitative confirmation that globalization did reach a turning point in the mid-2000s. This development was soon obscured by the Great Recession of 2008, but McKinsey's analysis of 23 industry value chains (interconnected activities, from design to retail, that deliver final products) spanning 43 countries between 1995 to 2017 shows that goods-producing value chains (still growing slowly in absolute terms) have become significantly less trade-intensive, with exports declining from 28.1 percent of gross output in 2007 to 22.5 percent in 2017.[96] What I see to be the study's second-most important finding is that, contrary to common perception, only about 18 percent of the global goods trade is now driven by lower labor costs (labor arbitrage), that in many chains this share has been declining throughout the 2010s, and that global value chains are becoming more knowledge-intensive and rely increasingly on highly skilled labor. Similarly, an OECD study shows that the expansion of global value chains stopped in 2011 and since then has slightly declined: there has been less trade in intermediate goods and services.[97]

Add to this the (justified or exaggerated, thoughtful or demagogic) fears about globalization's impact on national sovereignty, culture, and language; about diluting cherished peculiarities in the solvent of commercial universality (with concerns ranging from the ubiquity of American fast-food chains to the essentially unchecked power of social media); and, contrary to promised benefits, worries about globalization's role in economic and social inequality. Even a restrained appraisal of these real and perceived negatives confirms enough downsides to question any future intensification of the process, and in 2020 COVID-19 reinforced such sentiments.

Arguments for reshoring many kinds of manufacturing in order to gain greater resilience and reduce unexpected disruptions are not new. The progress of globalization and the actions of multinational companies have been questioned and criticized since the 1990s and, more recently, these sentiments became part of electoral discontent in some countries, most notably in the UK and the US.[98] But as the COVID-19 pandemic unfolded, a remarkable lineup of institutions began to publish analyses and appeals for the reorganization of global supply chains. The OECD looked at the policy options to build more resilient production networks that would rely less on imports from distant places and that could better withstand global trade interruptions. The United Nations Conference on Trade and Development considered repatriating manufacturing from Asia to North America and Europe and a shift to shorter, less fragmented value chains—extending from design through manufacturing to distribution within a single country or a single economic unit—that would produce a higher concentration of value added. Swiss Re produced a report about the de-risking of global supply chains (rebalancing them to strengthen resilience). And the Brookings Institution saw the reshoring of advanced manufacturing as the best way to create good jobs.[99]

Questioning and criticizing globalization has gone beyond narrowly ideological arguments, and the COVID-19 pandemic provided additional powerful arguments based on irrefutable concerns about the state's fundamental role in protecting the lives of its citizens. That role is hard to play when 70 percent of the world's rubber gloves are made in a single factory, and when similar or even higher shares of not just other pieces of personal protective equipment but also of principal drug components and common medications (antibiotics, antihypertensive drugs) come from a very small number of suppliers in China and India.[100] Such dependence might fulfill an economist's dream of mass output at the lowest possible unit cost, but it makes for extremely irresponsible—if not criminal—governance when doctors and nurses have to face a pandemic without adequate PPE, when states dependent on foreign production engage in dismaying competition for limited supplies, and when patients around the world cannot renew their prescriptions because of the slowdowns or closures in Asian factories.

And security concerns created by excessive globalization go far beyond the health-care sector. Rising US imports of large Chinese transformers create worries about the availability of spare parts and about the potential for future grid destabilization, and there is little need to repeat the arguments about the much-publicized ban on Huawei's participation in the 5G networks of some Western nations.[101] Not surprisingly, the reshoring of manufacturing could be the wave of the future, both in North America and in Europe: a 2020 survey showed that 64 percent of American manufacturers said that reshoring is likely following the pandemic.[102]

Will this sentiment persist? As I never fail to stress, I do not forecast, and hence I am not offering any specific numbers concerning the retreat or continuation of the pre-COVID levels of globalization in general, or of the reshoring of manufacturing capacities in particular. I just try to appraise the range of the most likely outcomes, and while in recent years it has looked increasingly as if most aspects of globalization will not soar to new highs, in 2020 this notion became entirely unexceptional: we may have seen the peak of globalization, and its ebb may last not just for years but for decades to come.

5. Understanding Risks

From Viruses to Diets to Solar Flares

One sweeping, simplifying way to describe the advances of modern civilization is to see them as serial quests to reduce the risks that come from us being complex and fragile organisms trying to survive against many odds in a world abounding with dangers. The previous chapters have documented how successful we have been in this quest. Higher crop yields have improved food supply, lowered its costs, and reduced the risks of malnutrition, stunting, and childhood diseases stemming from undernutrition. Most notably, the combination of expanded food production, extensive food trade, and emergency food aid have eliminated the long-standing inevitability of recurrent famines.[1]

Better housing (more space, running and hot water, central heating), better hygiene (no single improvement being more important than more soap and the more frequent washing of hands), and better public health measures (ranging from mass-scale vaccinations to food safety oversight) have improved domestic comfort, reduced the risks of infections spread from contaminated water, cut the frequency of foodborne pathogens, and largely eliminated the dangers of carbon monoxide poisoning from wood-fired stoves.[2] Various engineering advances and public safety measures reduced industrial and transportation accidents. Car accidents (with fatalities now in excess of 1.2 million a year) would be much more deadly without the improvements in car design and protective features (anti-intrusion bars guarding against side impacts, seat belts, airbags, brake lights at the driver's eye level, and, increasingly, automatic braking and lane-departure correction) that have cut the risks of collision and serious injury.[3]

International treaties set down transparent rules that promote reliability and safety (such as reducing the risks of importing contaminated goods) and that make regrettable events subject to legal action (such as pursuing a parent who abducted a child to another country).[4] And despite the impression created by media reporting, the worldwide frequency of violent conflicts and the total number of their casualties have been declining for decades.[5] But given the complexity of our bodies, the enormity and unpredictability of natural processes, and the impossibility of eradicating all human errors made when designing intricate machines and when operating them, it is not at all surprising that risks continue to abound in the modern world.

Even people who take no particular steps to be well informed are routinely exposed to media reports of man-made and natural dangers and the risks of diets, diseases, and quotidian activities. The first category ranges from dreaded terrorist attacks to many manifestations of chemophobia (from pesticide residues in food to carcinogens in toys or carpets), and from asbestos hidden in walls and in baby powder to the planet being ruined by anthropogenic global warming.[6] Media reports do not miss any news of natural catastrophes—including hurricanes, tornadoes, floods, droughts, and locusts—and in the background there are lasting worries about incurable cancers and unpredictable viruses, with recent concerns about SARS-CoV-1 and Ebola just mild previews of the anguish brought by the COVID-19 (SARS-CoV-2) pandemic.[7]

The list can be easily extended by adding worries about mad cow disease (bovine spongiform encephalopathy), *Salmonella* or *Escherichia coli*, exposure to hospital microbes (nosocomial infections), non-ionizing radiation from mobile phones, cybersecurity and data theft, artificial intelligence designs or genetically engineered organisms getting out of control, the accidental launch of nuclear missiles, and a stray, unobserved asteroid hitting the planet. With such a recital we might easily conclude that we are now exposed to more risks than ever—or, in contrast, that the incessant (and exaggerated) reporting of such events or their possibilities has simply made us more aware of their existence and that proper risk perception would provide some calming perspectives. And that is exactly what I will do in this chapter. Yes, the world is full of constant or episodic risks, but it is also replete with wrong perceptions and irrational risk appraisals. There are many reasons for these misperceptions and miscalculations, and practitioners of risk analysis have published revealing findings regarding their origins, prevalence, and endurance.[8]

But before I get to the analyses, quantifications, and comparison of man-made and natural risks, let's start with the basics. What should we eat to promote a long life? Given the veritable minefield of modern dietary claims and counterclaims, this might seem to be an impossible, or at least very difficult, question to answer. How will I weigh the respective merits and demerits of diets ranging from unchecked carnivory to the purest veganism? The first one, promoted as the supposed Paleolithic diet, supplies more than a third of all food energy from meat protein; the other extends beyond never swallowing even a microgram of animal matter to never wearing leather shoes, knitted woolen sweaters, or silk blouses. The first appeals to a caricature of some of our distant evolutionary roots; the second offers the surest path to the preservation of the long-suffering biosphere because humble plants, unlike destructive domesticated animals, exert only the gentlest pressure on the environment.[9]

My approach to finding the least risky diets (those associated with life expectations above 80 years) will ignore not only all of the dubious dietary claims promoted by the media, but also, perhaps more surprisingly, scores of publications in scientific journals. In particular, those that have examined the links between diets, diseases, and longevity by following groups of various sizes and ages for shorter or longer periods of time while relying overwhelmingly on the participants' recollections of all the food they have eaten in the past. I will also disregard metastudies of such projects. Merely listing these post-1950 publications—from the examination of coronary heart disease and saturated fat and cholesterol, to the risks of eating meat and drinking milk—would fill a small book, and a busy segment of these inquiries has been devoted to exposing the fallibility of human memory (what did you eat last week? I bet you can't remember, or at least not accurately), as well as detailing other methodological or analytical shortcomings, such that this field is rife with accusations of invalid conclusions.[10]

No wonder most people find the question of what we should eat a difficult one. These studies, and their metastudies, have repeatedly failed to produce consistent clear-cut outcomes, with new research often upending previous findings.[11] Is there a better way out of these generations-long, and still unfolding, dietary conundrums? Indeed, it is quite simple. We can look at which populations live the longest and what their diets are.

Eating as in Kyoto—or as in Barcelona

Among the world's more than 200 nations and territories, Japan has had the highest average longevity since the early 1980s, when its combined (male and female) life expectancy at birth surpassed 77 years.[12] Further gains followed, and by 2020 Japan's combined life expectancy at birth was about 84.6 years. Women live longer in all societies, and by 2020 their life expectancy in Japan was about 87.7 years, ahead of 86.2 years in second-place Spain. Average longevity is an outcome of complex and interacting genetic, lifestyle, and nutritional factors. Trying to find out to what extent it is determined by diet alone is impossible, but if there are unique features to a nation's diet, they clearly deserve a closer examination.

Is there anything truly special about Japan's food consumption that would provide a ready explanation of that diet's contribution to the nation's record longevity? All of its traditional ingredients consumed in substantial quantities differ only subtly from those eaten or drunk in abundance in neighboring Asian nations. The Chinese and Japanese consume different, but nutritionally equivalent, varieties of the same subspecies of rice (*Oryza sativa japonica*). The Chinese have traditionally coagulated their bean curd (*dòufu*) with calcium sulfate (*shígāo*), while the Japanese bean curd (*tōfu*) is gelled with magnesium sulfate (*nigari*), but the ground legume grain is identically rich in protein. And unlike unfermented Japanese green tea (*ocha*), Chinese green tea (*lǜchá*) is partially fermented. These are not differences in nutritional quality, merely matters of appearance, color, and taste.

The Japanese diet has undergone an enormous transformation during the past 150 years. The traditional diet, consumed by most of the nation before 1900, was insufficient to support the population's growth potential and it resulted in short statures among both women and men; slow pre–Second World War improvements accelerated after the country overcame food shortages following its defeat in 1945.[13] The consumption of milk, first introduced at school lunches to prevent malnutrition, began to rise, and white rice became abundant. Seafood supply expanded rapidly as the country built the world's largest fishing (and whaling) fleet. Meat became a part of common Japanese dishes, and many baked goods emerged as favorites in this traditionally non-baking culture. Higher incomes and hybridization of tastes brought increases in mean blood cholesterol levels, blood pressure, and body weight—and yet heart disease did not soar and longevity increased.[14]

The latest published surveys show Japan and the US to be surprisingly close in total food energy consumed per day. In 2015–2016, US males consumed only 11 percent more, and US women not even 4 percent more food energy per day than their Japanese counterparts did in 2017. The two countries diverged moderately in total carbohydrate (Japan was ahead by less than 10 percent) and protein (with Americans less than 14 percent ahead) consumption, and both nations were well above the needed protein minima. But there is a major gap in terms of average fat intake, with American males consuming about 45 percent more and women 30 percent more than the Japanese. And the greatest disparity is in sugar intake: among US adults it is about 70 percent higher. When recalculated in terms of average annual differences, Americans have recently consumed about 8 kilograms more fat and 16 kilograms more sugar every year than the average adult in Japan.[15]

The widespread availability of ingredients, and easy access to cooking instructions and recipes on the internet, means that you too can minimize your risk of premature mortality and start eating *à la japonaise*—be it the country's traditional cuisine, *washoku*, or its adaptations of foreign meals (Wienerschnitzel appearing as pre-sliced *tonkatsu*; curry and rice transformed into gooey *kare raisu*).[16] But before you start breakfasting on miso soup (*miso shiru*), lunching on plain cold *onigiri* (rice balls wrapped in *nori*, dried seaweed), and dining on *sukiyaki* (meat and vegetable stew) a second opinion might be in order: what would the best European model of diet and longevity do?

The Spanish women are the runners-up in the world's record life expectancy, and the country traditionally followed the so-called Mediterranean diet, with high intakes of vegetables, fruits, and whole grains complemented by beans, nuts, seeds, and olive oil. But as the average incomes in Spain rose, they rapidly changed those habits to a surprisingly high degree.[17] Until the late 1950s, Franco's impoverished Spain continued to eat very frugally. Typical diets were dominated by starches (annual consumption of cereals and potatoes added up to about 250 kilograms per capita) and vegetables; meat supply (carcass weight) remained below 20 kilograms per capita and actual consumption was less than 12 kilograms (of which a third was mutton and goat meat); *aceite de oliva* was the most important plant oil (about 10 liters a year); and only sugar consumption (about 16 kilograms in 1960) was high in relation to other foodstuffs.

Dietary changes accelerated after Spain joined the EU in 1986, and by the year 2000 it became Europe's leading carnivorous nation (after more than quintupling the average per capita supply to just above 110 kilograms per year). A subsequent slight decline cut the rate (carcass weight) to about 100 kilograms per capita in 2020, but that is still twice the Japanese mean! And with dairy products and cheeses added to fresh meat and the enormous quantity and variety of *jamones* (hams cured by salting and prolonged drying), it is hardly surprising that the Spanish supply of animal fat is four times the Japanese rate.[18] Spaniards now consume almost twice the volume of plant oils as the Japanese—but their consumption of olive oil is about 25 percent lower than in 1960.

Higher incomes have only increased the traditional predilection for sugary creations, and the adoption of pop drinks did the rest: since 1960, per capita consumption of sugar has doubled and it is now about 40 percent above the Japanese level. At the same time, Spanish wine-drinking has been relentlessly declining, from about 45 liters per capita in 1960 to just 11 liters by 2020, and beer has become by far the country's most-consumed alcoholic drink. The way Spain now eats is substantially different from the way Japan feeds itself—and, most definitely (being the continent's top carnivore), this diet hardly resembles the frugal, near-vegetarian, and life-lengthening legendary Mediterranean diet.

But despite a more meaty, fatty, and sugary diet (and also rapidly abandoning drinking its supposedly heart-protecting wines), Spain's cardiovascular mortality has kept on declining and life expectancy has been rising. Since 1960, Spain's CVD mortality has been falling at a faster pace than the average of affluent economies, and by 2011 it was about a third lower than their mean; and since 1960, Spain has added more than 13 years to its combined (male and female) longevity prospect, raising it from 70 to more than 83 years by 2020.[19] This is just one year less than in Japan: is one additional year of life (and there are high chances that it might be spent in physical or mental decrepitude, or a combination of both) worth replacing half of the meat you eat with tofu?

Think of what you might be missing: those paper-thin slices of *jamón ibérico*; that well-roasted pig (even if not done as famously as at Sobrino de Botín, a short walk south of the Plaza Mayor, where they have been preparing it for nearly 300 years); that well-cooked *polpo gallego*, octopus stewed with potatoes, olive oil, and paprika. These are, truly, existential decisions to make—but the conclusion is reasonably clear. If we were to stake longevity (accompanied by healthy and active life) solely on the prevailing diet—which, however important, is but one element of a bigger picture that includes your inherited genes and surrounding environment—then Japanese eating has a slight edge, but an only slightly inferior outcome can be had by eating as they do in Valencia.

This is a vastly consequential but relatively simple risk assessment: one choice, based on convincing data, can suffice for decades to come. Other risk assessments are invariably trickier, where metrics might not be as simple as years lived. The risks of specific activities change over time (driving in the US is now generally much safer than half a century ago, but after 50 years of driving your skills might have deteriorated and you pose a greater risk to yourself and others when you get behind the wheel). And if you want to know if intercontinental flying (which you might do infrequently) is riskier than downhill skiing (which you may have done for many years), you must have a rather accurate comparative yardstick. And how does one compare the risks experienced in different nations—say, driving in the US, being struck by lightning while hiking in the Alps and getting killed by an earthquake in Japan? As it turns out, we can do some remarkably accurate, comparative evaluations of all of these risks.

Risk perceptions and tolerances

In his pioneering 1969 analysis of risks, Chauncey Starr—at that time the dean of the School of Engineering and Applied Science at the University of California in Los Angeles—stressed the major difference in risk tolerance between voluntary and involuntary activities.[20] When people think that they are in control (a perception that may be incorrect but that is based on previous experiences and hence on the belief that they can assess the likely outcome), they engage in activities—climbing vertical rock faces without ropes, skydiving, bullfighting—whose risks of serious injury or fatality may be a thousand-fold higher than the risk associated with such dreaded involuntary exposure as a terrorist attack in a large Western city. And most people have no problem engaging daily and repeatedly in activities that temporarily increase their risk by significant margins: hundreds of millions of people drive every day (and many apparently like to do so), and an even higher risk is tolerated by an even larger number of smokers[21]—in affluent countries, decades of education has reduced their ranks, but worldwide there are still more than 1 billion of them.

In some cases this disparity between tolerating voluntary risks and trying to avoid wrongly perceived risks of involuntary exposures becomes truly bizarre, as people refuse to have their children inoculated (voluntarily exposing them to multiple risks of preventable diseases) because they consider government requirements to protect their children (an involuntary imposition) as unacceptably risky—and have been doing so on the basis of repeatedly discredited "evidence" (most notably linking vaccination to a higher incidence of autism) or rumored perils (the implanting of microchips!).[22] And the SARS-CoV-2 pandemic elevated these irrational fears to a new level. Humanity's best hope to end the pandemic was mass-scale vaccination, but long before the first vaccines were approved for distribution, large shares of the population were telling pollsters that they would not get inoculated.[23]

Widespread fear of nuclear electricity generation is yet another excellent example of risk misperception. Many people smoke and drive and eat excessively but have reservations about living next to a nuclear power plant, and polling has shown lasting and pervasive distrust of this form of electricity generation despite the fact that it has prevented a large number of air pollution–related deaths that would have been associated with burning fossil fuels (by 2020, nearly three-fifths of the world's electricity came from fossil fuels, and just 10 percent from nuclear fission). And the comparison between overall risks of nuclear and fossil-fueled electricity generation does not flip even when the best estimates of all latent fatalities from the two major accidents (Chornobyl in 1985 and Fukushima in 2011) are included.[24]

Perhaps the most stunning contrast of nuclear-related risk perceptions is seen when comparing France and Germany. France has been deriving more than 70 percent of its electricity from nuclear fission since the 1980s and nearly 60 reactors dot the country's landscape, cooled by water from many French rivers, including the Seine, Rhine, Garonne, and Loire.[25] Yet the longevity of the French population (second only to Spain within the EU) is the best testimony to the fact that these nuclear power plants have not been a discernible source of ill health or premature deaths—but across the Rhine it is not only the German Greens who believe that nuclear power is an infernal invention that must be eliminated as fast as possible, but much larger portions of society too.[26]

This is why many researchers have argued that there is no "objective risk" waiting to be measured because our risk perceptions are inherently subjective, dependent on our understanding of specific dangers (familiar vs. new risks) and on cultural circumstances.[27] Their psychometric studies showed that specific hazards have their unique patterns of highly correlated qualities: involuntary risks are often associated with the dread of new, uncontrollable, and unknown hazards; voluntary hazards are more likely to be perceived as controllable and known to science. Nuclear electricity generation is widely perceived as unsafe, x-rays as tolerably risky.

Feelings of dread play an outsized role in risk perception. Terrorist attacks are perhaps the best example of this differentiated tolerance, as fear takes over and drives out rational assessment readily made on the basis of incontrovertible evidence. Because of their unpredictable timing, location, and scale, terrorist attacks rank high on the psychometric scale of dread, and these fears have been intensively exploited by vastly exaggerated pseudo-analyses offered by talking heads on 24/7 news channels: during the past two decades they have speculated on everything from suitcase-sized nuclear bombs detonated in mid-Manhattan to the poisoning of reservoirs used to supply drinking water to large cities and the spraying of deadly engineered viruses.

Compared to such dreaded attacks, driving presents largely voluntary, highly recurrent, and very familiar risks, and accidental deaths involve overwhelmingly (more than 90 percent of cases) only one person per fatal collision. As a result, societies tolerate the global toll exceeding 1.2 million deaths a year, something they would never assent to if it were to take the form of recurrent accidents in industrial plants or collapsed structures (bridges, buildings) in or near large cities, even if the combined annual death toll of such disasters was an order of magnitude smaller—"just" in the hundreds of thousands of fatalities.[28]

Large differences in individual tolerance of risk are best illustrated by the fact that many individuals engage—voluntarily and repeatedly—in activities that others might consider not just too risky but belonging all too clearly to the category of death wish. Base (fixed object) jumping is an excellent example of such an activity, as the slightest delay in opening the parachute may cost a life—a free-falling body reaches fatal velocity in a matter of seconds.[29] And then there is risk tolerance justified by fatalistic beliefs: diseases or accidents are predestined and inevitable, and hence it makes no sense to try enhancing one's health or preventing mishaps by appropriate personal action.[30]

Fatalistic people also underestimate risks in order to avoid the effort required to analyze them and draw practical conclusions, and because they feel totally unable to cope with them.[31] Traffic fatalism has been particularly well studied. Fatalistic drivers underestimate dangerous driving situations, are less likely to practice defensive driving (no distractions, keeping safe trailing distance, no speeding), and are less likely to restrain their children with seat belts or to report involvement in road accidents. Worryingly, studies in some countries found traffic fatalism prevalent among taxi drivers, and pervasive among minibus drivers.[32]

There is little we can do to convert base jumpers into paragons of risk-averse behavior or to convince many taxi drivers that their accidents are not predetermined. But we can use the best available understanding of risks, both of those in everyday life and those that are exceedingly uncommon but potentially deadly, to quantify their consequences and hence to compare their impacts. This is not an easy task, because we have to deal with such a variety of events and processes. Moreover, there is no perfect metric to do so, and there can be no universal yardstick to compare the ubiquitous risks faced daily by billions of individuals with the extraordinarily rare events that may take place just once in a hundred, a thousand, or even ten thousand years, but with catastrophic global consequences. In any case, that is what I will try to do.

Quantifying the risks of everyday life

For older people the danger starts even before waking up: heart attacks (acute myocardial infarctions) are more common, and more severe, during the dark-to-light transition period.[33] When they get up, one of the most common ways older people injure themselves is to fall. In the US, millions of accidental falls take place every year, leaving bruises or broken bones—and more than 36,000 deaths, disproportionately among those more than 70 years old, and often occurring not when walking up or down stairs but when simply losing balance or tripping over a carpet edge.[34] And once you make it to your kitchen there are food-associated risks, from *Salmonella* in improperly cooked eggs to pesticide residue in tea (this is a minuscule but, for drinkers of non-organic tea, everyday exposure).[35]

A morning drive may be on an icy road or a drugged-up driver could go through a red light. Your office walls may still hide old asbestos insulation, and faulty air conditioning may spread *Legionella* bacteria. Your coworkers may infect you with seasonal flu or (as happened in 2020–2021, 2009, 1968, and 1957) with a new pandemic virus. You may have a severe allergic reaction to a nut accidentally admixed into a nut-free chocolate bar. If it is tornado season in Texas or Oklahoma you might return from work to see your home turned into a pile of rubble, and if you live in Baltimore you cannot remain unconcerned about the city's homicide rate that is an order of magnitude higher than in Los Angeles, the city famous for its gangs.[36] And as hardly any generic drugs are made domestically (they come mostly from China and India), your drugstore may not fill your prescription because a contaminated batch was pulled from distribution.[37]

And detailed data on age- and sex-specific death rates show how the reasons for (and hence the concerns about) getting mortally sick change as people grow older. The latest statistics show that among males in England and Wales heart disease dominates from the early 50s to the late 70s, and for women breast cancer becomes the most dreaded disease by their mid-30s and it remains so until their mid-60s; afterwards, lung cancer is the single largest cause of death among females, and dementia and Alzheimer's disease have recently displaced ischemic heart disease as the leading cause of death for both sexes over 80 years of age.[38]

Quantifying common risks seems to be a daunting enterprise. How does one compare the risks of dying due to an unusually severe seasonal influenza epidemic to the risk of a mortal injury resulting from occasional weekend kayaking or snowmobiling; or the risk of frequent transpacific flying to the risk of habitual eating of California-grown lettuce that might be repeatedly contaminated with *Escherichia coli*? And how do we express fatal risks? Per standard number of people (1,000; 1 million) in an affected population? Per unit of hazardous substance, per unit of time exposure, or per unit of ambient concentration?

A uniform metric able to subsume fatalities and injuries or economic losses (whose totals could differ by orders of magnitude among different societies) and chronic pain (something that remains notoriously unquantifiable) is clearly an impossible objective. But the finality of dying provides a universal, ultimate, and incontestably quantifiable numerator that can be used for comparative risk assessment. The simplest and most obvious way to make some revealing comparisons is to use a standard denominator and to compare annual frequencies of causes of death per 100,000 people. When using the US statistics (the latest published detailed breakdown is for 2017) this leads to some surprising outcomes.[39]

Homicides take almost as many lives as leukemia (6 vs. 7.2), a dual testament to the advances in treating that malignancy and to the extraordinary violence of American society. Accidental falls kill almost as many people as the dreaded pancreatic cancer with its short post-diagnosis survival (11.2 vs. 13.5). Motor vehicle accidents take twice as many lives (and, moreover, much younger ones) than does diabetes (52.2 vs. 25.7), and accidental poisoning and noxious substances exact a higher death toll than does breast cancer (19.9 vs. 13.1). But these comparisons use the same denominator (100,000 people) without taking into account the duration of exposure to a given cause of death. Homicides can, and do, take place in public and private and at any time of day or night, and the exposure to this risk is thus 24 hours a day, 365 days a year—but motor vehicle accidents (including those that kill pedestrians) can happen only when somebody is driving, and most Americans spend only about an hour behind the wheel every day.

A more insightful metric then is to use the time during which people are affected by a given risk as the common denominator, and do the comparisons in terms of fatalities per person per hour of exposure—that is, the time when an individual is subject, involuntarily or voluntarily, to a specific risk. This approach was introduced in 1969 by Chauncey Starr in his evaluation of social benefits and technological risks and I still find it preferable to another general metric—that of micromorts.[40] These units define a micro probability, a one-in-a-million chance of death per specific exposure, and express it per year, per day, per surgery, per flight, or per distance traveled—and these non-uniform denominators do not make for easy across-the-board comparisons.

Overall death rates (per 1,000 people) are well monitored worldwide, both for populations at large and for each sex by specific age group.[41] Overall mortality depends heavily on the population's average age. In 2019 the global mean was 7.6/1,000, while Kenya's mortality (despite a lower standard of nutrition and health care) was less than half of the German rate (5.4 vs. 11.3) because Kenya's median age of just 20 years is less than half of Germany's 47 years. Data on deaths due to specific diseases are also commonly available—with cardiovascular diseases accounting for a quarter of the total in the US (2.5/1,000) and cancers for a fifth (2/1,000)—as is the information on deaths due to injuries (ranging from about 1.4 for falls and 1.1 for transport accidents, to 0.7 for encounters with animals and just 0.03 for accidental poisonings) and natural disasters.[42]

The entire year (8,766 hours when corrected for leap years) is the denominator for overall mortality, for chronic diseases, and for natural disasters such as earthquakes or volcanic eruptions that can strike anytime. But in order to calculate risks for such common activities as driving or flying, we have to ascertain first the totals of specific populations engaged in these activities and then to estimate the average hours of annual exposure. The same sequence applies to quantifying the risks of dying in a hurricane or a tornado: these cyclones are not around every day of the year and do not affect the entirety of large countries.

Calculating the baseline, the average population-wide or sex- and age-specific risk of overall mortality, is easy. In 2019, overall mortality (crude death rate) of well-off (developed) countries clustered at around 10/1,000, with actual rates ranging from 8.7 for North America to 10.7 in Japan and 11.1 for Europe. That annual mortality of 10/1,000 (with 1,000 people subject to dying for 8,766 × 1,000 hours) prorates to 0.000001 or 1×10^{-6} per person per hour of exposure. Cardiovascular diseases are the leading cause of mortality in all affluent countries and they account for nearly a quarter of that total (3×10^{-7}). Seasonal influenza carries a risk an order of magnitude lower (usually about 2×10^{-8} and up to 3×10^{-8}), and even in the violence-prone United States the risk of homicide has recently been just 7×10^{-9} per hour of exposure, half the risk of death attributable to falls (1.4×10^{-8}). But, as already noted, the frequency of the latter kind of accidental death is highly skewed, with people over 85 years of age having a risk of 3×10^{-7} compared to just 9×10^{-10} for people 25–34 years old.[43]

To reverse the conclusion about general mortality, in affluent countries the overall risk of natural demise amounts to 1 person among 1 million dying every hour; every hour, 1 person among about 3 million dies of heart disease and 1 among roughly 70 million dies of an accidental fall. Such odds are sufficiently low not to preoccupy an average citizen of any affluent country. Sex- and age-specific numbers are, inevitably, different. While Canada's overall mortality for both sexes is 7.7/1,000, for young (20–24 years) males it is only 0.8/1,000 but for men of my age (75–79 years) it is 35/1,000, and my group's risk is then 4×10^{-6} per person per every hour of being alive, four times the rate for the population average.[44]

Before I turn to quantifying the risks of voluntary activities, I should clarify the dangers associated with hospital stays. They are unavoidable because of many conditions (and in many countries also increasingly for elective cosmetic surgeries), and high patient throughputs make it more likely that medical errors will happen. In 1999, the first study of preventable medical errors found that anywhere between 44,000 and 98,000 of them are made in the US every year.[45] That was an uncomfortably high total—and in 2016 a new study raised it to 251,454 cases in 2013 (and possibly to as many as 400,000 deaths), making it the third-highest cause of US mortality in that year, behind heart disease (611,000) and cancers (585,000) and ahead of chronic respiratory disease (149,000).[46] These results, widely reported in the mass media, implied that every year 35–58 percent of all hospital deaths in the country were due to medical error.

When put in this way, the implausibility of those claims become easily evident: careless errors surely take place, regrettable omissions do happen, but that they could add up to anywhere between just over a third and nearly three-fifths of all hospital deaths would brand modern medicine as an extraordinarily inept, if not outright criminal, endeavor. Fortunately, these high mortalities did not result from carelessness but from data-handling errors.[47] The latest study of mortality associated with adverse effects of medical treatment (AEMT) sets the record straight: it found 123,063 such deaths between 1990 and 2016 (mostly due to surgical and perioperative errors), a decline of 21.4 percent to 1.15 AEMT deaths per 100,000 people.[48]

Men and women had similar rates, but states differed significantly, with California as low as 0.84 AEMT deaths per 100,000. In absolute terms this averages to about 4,750 deaths a year, less than 2 percent of the lowest estimate published in 2016.[49] Translated into a comparative risk metric, this results in about 1.2×10^{-6} fatalities per hour of exposure, which means that any elderly male reader of this book (whose general mortality risk is between 3×10^{-6} and 5×10^{-6}) will increase his risk of demise due to AEMT by no more than about 20–30 percent during the few days of an average stay in an American hospital—and that, I would argue, is a very encouraging risk finding!

Voluntary and involuntary risks

How much do we increase these baseline risks, or risks associated with such unavoidable events as emergency operations or short hospital stays required for medical evaluations, by voluntary exposures through participating in a wide variety of more or less risky endeavors? And how much should we worry about unavoidable involuntary risk resulting from natural hazards ranging from earthquakes to floods?

As already noted, these are useful categories for risk assessment, but the distinction between voluntary and involuntary exposures is not always obvious. There are clear-cut voluntary (and fairly to very risky) activities such as smoking or engaging in extreme sports; and obviously unavoidable involuntary risks both at an individual level (including the exceedingly low danger of being hit by a meteorite) and as collective, indeed a planet-wide, experiences (the Earth colliding with an asteroid being the foremost example).

But many risky exposures cannot be so easily assigned, because there is no clear dichotomy between voluntary and involuntary risks: driving to work may be a matter of choice for a family that built a dream exurban house, but it is a matter of unavoidable necessity for millions of people in North America with its notoriously poor mass transit systems. And if a young man wants to stay in Newfoundland, there are not that many choices beyond becoming a fisherman or a worker on a massive oil-producing platform, both being far more risky occupations than moving to Toronto, learning how to code, and writing apps in a glassed office far away from the rock jutting into the North Atlantic.

Keeping these complications in mind, I will first explain the risks associated with driving and flying, activities that globally involve hundreds of millions of vehicle drivers and passengers and recently more than 10 million paying fliers every day. For both activities, we must start by accurately counting the number of deaths and then deploying necessary assumptions in order to define affected populations and their aggregate time of exposure to a given risk.

For driving it is obviously time spent behind the wheel (or as a passenger). For the US we have totals of distances traveled every year by all motor vehicles and by passenger cars (a recent grand total has been about 5.2 trillion kilometers annually) and, after declining for many years, traffic fatalities have gone up slightly to about 40,000 a year.[50] In order to estimate the time spent driving, we have to divide the distance driven by average speed—and, obviously, this number can be only a defensible approximation, not an accurate rate. Intercity speeds show less variation, but urban speeds tend to drop by as much as 40 percent during recurrent rush hours. Assuming an average combined speed of 65 km/hour (about 40 mph) gives us annually about 80 billion driving hours in the US, and with 40,000 fatalities this translates exactly to 5×10^{-7} (0.0000005) fatalities per hour of exposure. Neither the fact that traffic fatalities also include pedestrians and bystanders killed by vehicles nor the deployment of other plausible average speeds (say, 50 or 70 km/hour) would change the order of magnitude. Driving is an order of magnitude more dangerous than flying, and during the time a person is driving the average chance of dying goes up by about 50 percent compared to staying at home or tending a garden (as long as that does not include climbing a tall ladder or working with a large chainsaw).

And for men of my age group the driving-risk bump is only 12 percent above the overall risk of dying. Driving risks in the US also show significant differences due to gender and population groups. The lifetime risk of dying in a motor vehicle accident is only 0.34 percent for Asian American females (1 out of 291) but 1.75 percent (1 of every 57) for Native American males, while the risk for all individuals is 0.92 percent (1 out of 109).[51] Of course, in other countries where people drive much less than Americans and Canadians but where the accident rates are much higher (they are about twice as common in Brazil, three times as much in sub-Saharan Africa), the risks are up to an order of magnitude greater.[52]

Scheduled commercial flights, already a very low-risk activity at the end of the last century, got appreciably safer during the first two decades of the 21st century. This conclusion stands despite some disturbing recent losses, including the still-unsolved (and likely never to be explained) disappearance of Malaysia Airlines flight 370 somewhere over the Indian Ocean in March 2014, followed by the downing of Malaysia Airlines flight 17 over eastern Ukraine in July 2014, and the two crashes of the new Boeing 737 MAX—Lion Air flight 610 in the Java Sea (October 29, 2018) and Ethiopian Airlines flight 302 near Addis Ababa (March 10, 2019).[53]

Perhaps the most revealing way to compare the airline industry's fatalities is per 100 billion passenger-kilometers flown. This rate was 14.3 in 2010, it reached a record low of 0.65 in 2017, but it increased to 2.75 in 2019. Flying in 2019 was thus more than five times safer than in 2010, and more than 200 times safer than at the beginning of the jetliner era in the late 1950s.[54] Expressing these fatalities in terms of risks per hour of exposure is fairly straightforward. The mean 2015–2019 total of accidental deaths was 292; the averages of 68 trillion passenger-kilometers flown and 4.2 billion passengers mean that average passengers flew about 1,900 kilometers and spent about 2.5

hours in flight; the total of about 10.5 billion passenger-hours spent aloft and 292 fatalities translates to 2.8×10^{-8} (0.000000028) fatalities per person per hour of flying. This is only about 3 percent of the general risk of mortality while aloft, and in the case of a septuagenarian male the risk while aloft rises by a mere 1 percent. Any rational frequent flyer (and even more so an elderly one) should worry more about encountering unforeseen delays, running the security theater gauntlet, enduring the tedium of long-distance flying, and coping with the debilitating effects of jetlag.

At the opposite end of the voluntary risk spectrum are activities whose brief duration carries a high probability of death. None is riskier than base jumping from cliffs, towers, bridges, and buildings. The most reliable study of this "asking for it" madness looked at an 11-year period of jumping from the Kjerag Massif in Norway, where 1 in every 2,317 jumps (9 in total) resulted in death,[55] with an average exposure risk of 4×10^{-2} (0.04). For comparison, in skydiving a fatal accident used to take place roughly once every 100,000 jumps but the latest US data show one fatality for every 250,000 jumps. With a typical descent lasting five minutes the exposure risk is only about 5×10^{-5}, still 50 times higher than just sitting in a chair for those five minutes—but it is only about 1/1,000 of the risk associated with base jumping.[56] Again, only very few people are actually aware of these specific numbers, but nearly all people (save for the risk-tolerant few) behave as if they have internalized them.

In 2020 in the US, about 230 million people held a driver's license (exposure risk behind the wheel is 5×10^{-7} per person per hour); about 12 million were downhill skiers (2×10^{-7} while descending); the United States Parachuting Association has about 35,000 members (5×10^{-5} while aloft); the US Hang Gliding & Paragliding Association has about 3,000 members and what they do (depending on the length of flights lasting anywhere between 20 minutes and a few hours) carries a fatality risk of 10^{-4} to 10^{-3}; and although base jumping has been increasing in popularity (particularly in Norway and Switzerland), in the US it is still confined to a few hundred mostly male fate-tempters whose risk of dying is 4×10^{-2} during their brief falls.[57] The strong inverse relationship between the risk and overall participation in an activity is obvious:

large numbers of people are willing to risk a dislocated shoulder or a sprained ankle while skiing downhill on a groomed run; very few are into launching themselves into the void from precipices.

Finally, a few key numbers concerning one of the most dreaded modern involuntary exposures: the risk of terrorism. Between 1995 and 2017, 3,516 people died in terrorist attacks on US soil, with 2,996 fatalities (or 85 percent of that total) on September 11, 2001.[58] Countrywide individual exposure risk thus averaged 6×10^{-11} during those 22 years, and for Manhattan it was two orders of magnitude higher, increasing the risk of just being alive by one-tenth of a percent, a quantity that is too small to be meaningfully internalized. In less fortunate countries, the recent toll of terrorist attacks has been much higher: in Iraq in 2017 (with more than 4,300 deaths) the risk rose to 1.3×10^{-8}, and in Afghanistan in 2018 (7,379 deaths) to 2.3×10^{-8}, but even that rate raises the basic risk of being alive by just a few percent and it remains lower than the risk people voluntarily assume by driving (particularly in places with no lanes and ad hoc traffic rules).[59]

But correct as they are, these comparisons also show the inherent limits of dispassionate quantification. Most people commuting to work by car drive only at specific times, spend rarely more than an hour or an hour and a half per day on the road, follow familiar routes, and (save for inclement weather or an unexpected traffic jam) feel pretty much in control. In contrast, during times of peak terror, bombings or shooting attacks in Kabul or Baghdad took place at unpredictable times and intervals, in many public places—from mosques to markets—and there is no reliable way to completely avoid such threats while living in a city. As a result, the lower exposure rates to terrorist threats carry an unquantifiable accompaniment of dread, qualitatively so different from being concerned about possibly slippery roads during a morning commute.

Natural hazards: less risky than they look on TV

And how do recurrent deadly natural hazards compare with just being alive, and with the risks of extreme sports? Some countries are repeatedly (but not very frequently) subject to just one or two kinds of catastrophic events—flooding and extremely strong winds in the case of the UK—while the US has to cope every year with many tornadoes and extensive flooding, frequently with hurricanes (since the year 2000, nearly two hurricanes a year made a landfall) and heavy snowfalls, and its Pacific states are always at risk of experiencing a major earthquake and possible tsunami.[60]

Tornadoes kill people and destroy homes every year, and detailed historical statistics make it possible to calculate accurate exposure risks. Between 1984 and 2017, 1,994 people were killed in the 21 states with the highest frequency of these destructive cyclones (the region between North Dakota, Texas, Georgia, and Michigan, with about 120 million people), and about 80 percent of those deaths took place in the six months of the year from March to August.[61]

This translates to about 3×10^{-9} (0.000000003) fatalities per hour of exposure, a risk that is three orders of magnitude lower than just living. Very few inhabitants of America's tornado-swept states are aware of this rate but they recognize—as do people in other areas subject to recurrent natural catastrophes—that the probability of being killed by a tornado is sufficiently small, and hence the risk of continued living in such regions remains acceptable. Widely broadcast images of destruction left by powerful tornadoes make viewers living in atmospherically less violent regions wonder why people say that they will rebuild on the same spot. But such decisions are neither irrational nor recklessly risky, and because of them millions of people continue to live in the Tornado Alley that extends from Texas to South Dakota.

Remarkably, calculations of exposure risks to other commonly encountered natural disasters around the world converge on the same order of magnitude (10^{-9}) or yield even lower rates. Again, these low average fatality exposure rates help to explain why entire countries come to terms with the ever-present risks of earthquakes. Between 1945 and 2020, Japanese earthquakes (which can affect every part of the island nation) killed about 33,000 people, more than half as a result of the March 11, 2011 Tōhoku earthquake and tsunami (15,899 deaths and 2,529 missing).[62] But for a population that grew from 71 million in 1945 to nearly 127 million in 2020, that works out to about

5×10^{-10} (0.0000000005) fatalities per hour of exposure, four orders of magnitude lower than the country's overall mortality rate: obviously adding 0.0001 to 1 can hardly be a decisive factor that changes the overall assessment of life's risks.

Floods and earthquakes in most parts of the world carry exposure risks mostly on the order of between 1×10^{-10} and 5×10^{-10}, and the post-1960 rate for American hurricanes (potentially affecting about 50 million people in the coastal states from Texas to Maine and killing, on average, about 50 people a year) has been about 8×10^{-11}.[63] That is a remarkably low rate—either very similar to or perhaps even lower than what most people would consider an exceptionally low natural risk: being killed by lightning. In recent years, lightning has killed fewer than 30 people a year in the US, and when assuming that the danger applies only when outdoors (averaging four hours a day) and during the six months from April to September (when about 90 percent of all lightning occurs) the risk equals about 1×10^{-10}, while extending the exposure period to 10 months lowers it to 7×10^{-11} (0.00000000007).[64]

The fact that US hurricanes now present a fatality risk no greater than lightning illustrates how their toll has been reduced by satellites, advanced public warnings, and evacuations. At the same time, there are reasons for concern, as both the annual worldwide frequency of natural disasters and their economic cost have been increasing. We can say this with a high degree of confidence because the world's largest reinsurance companies (whose profits and losses depend on the unpredictable occurrences of earthquakes, hurricanes, floods, and fires) have been carefully monitoring their trends for decades.

Insurance is an ancient practice of providing different degrees of compensation for a variety of risks. While life insurance is based on highly predictable survival rates, insuring against unpredictable major natural hazards forces insurance companies to share the risk associated with such disasters by securing their own insurance. As a result, the world's largest reinsurance companies (Swiss Re, German Munich Re and Hannover Rueck, French SCOR, US Berkshire Hathaway, British Lloyd's) are the most assiduous students of natural catastrophes because their very existence depends on making proper calls: in order to avoid rising insured losses, they should not set their insurance premiums on the basis of outdated numbers that would underestimate future risks.

Numbers of all natural catastrophes recorded by Munich Re show expected year-to-year fluctuations but the upward trend has been unmistakable: a slow increase between 1950 and 1980, a doubling of annual frequency between 1980 and 2005, and about a 60 percent rise between 2005 and 2019.[65] Overall economic losses (reflecting exceptional burdens stemming from major disasters) show even greater annual fluctuations and an even steeper rising trend. When measured in constant 2019 monies, the pre-1990 record was about $100 billion, while 2011 set an all-time record of just over $350 billion and that total was nearly matched in 2017. Insured losses ranged mostly between 30 and 50 percent of total losses, with the 2017 record reaching nearly $150 billion.

Until the 1980s, the rising disaster toll was mainly attributable to greater exposure (resulting from growing populations and economies), and while this trend continues—there are more people with more insured property living in disaster-prone regions—recent decades have seen changes in natural hazards themselves: a warmer atmosphere holds more water vapor (increasing the chances of extreme precipitation); prolonged droughts in some regions cause recurrent fires of exceptional duration and intensity. Many models now forecast further intensification of these trends, but we also know that many effective measures—from setting up exclusion zones and restoring wetlands to enacting proper building codes—can be taken to reduce their impacts.

In order to get even lower risks from exposures to natural or man-made hazards, one must hunt for truly exceptional events such as people killed by a falling meteorite or by debris from an increasing number of orbiting satellites. A report by the US National Research Council estimated that, given the quantity of space debris hitting the Earth, there should be 91 deaths a year—and that would imply about 1×10^{-12} fatalities per hour of exposure for the global population of 7.75 billion. In reality, there have been no recorded deaths since 1900, and only recently the first written proof of a meteorite killing a man (and leaving another paralyzed) was discovered among the manuscripts in the General Directorate of State Archives of the Ottoman Empire: the event took place on August 22, 1888, in what is now Sulaymaniyah in Iraq.[66] But even if one person were killed every year, the rate would be merely 10^{-14}—or eight orders of magnitude smaller (1/100,000,000 as large) than just being alive, so clearly not a reason to worry.[67] As for the orbiting space junk, by 2019 there were some 34,000 pieces larger than 10 centimeters and more than 25 times as many pieces measuring 1–10 centimeters. All of these pieces break up as they re-enter the atmosphere, but even small pieces present collision risks in increasingly congested prime-orbit space.[68]

Ending our civilization

When we think about rare but truly extraordinary risks that have global effects, and even more so when we contemplate catastrophic events that could severely damage or even end modern civilization, we do so on an altogether different mental plane: those real (albeit very low) risks belong to a very different perception category. As with every event that might take place in a possibly quite distant future, we strongly discount their impact and, as demonstrated yet again by the 2020 pandemic, we are chronically unprepared to deal even with those risks whose recurrence is measured in decades, not in centuries or millennia.

Risks with truly global impacts fall into two very different categories: relatively frequent viral pandemics that can exact a considerable toll in a matter of months or a few years; and exceedingly rare but uncommonly deadly natural catastrophes that could take place within spans as short as a few days, hours, or seconds but whose consequences might persist not only for centuries but for millions of years, far beyond any civilizational horizons. Should a nearby supernova explode and flood the Earth with lethal doses of radiation from cosmic rays, would we have enough time (between the arrival of light and radiation) to improvise shelters for most of the global population?[69] But should this be a worry at all?

A blast that would damage the Earth's ozone layer must take place less than 50 light-years away, but all of our "nearby" stars that might possibly explode are much farther than this, and while a gamma-ray burst could affect Earth from as far as 10,000 light-years once every 15 million years, the closest such burst on record was 1.3 billion light-years away.[70] Clearly, this risk belongs to a largely academic category—rather than guessing when it might happen, we should, given the frequency of such events, better ask: will any terrestrial civilization be around in, say, 150,000 or half a million years? Although it is a comparatively more probable event, calculating the risk of an inevitable future collision of the Earth with an asteroid is another exercise in uncertainties and in assumptions whose particularities can make an enormous difference. Encounters with asteroids or large comets happened in the past and they will happen in the future—but do we assume that a major encounter takes place once every 100,000 years or once every 2 million years?[71]

These are relatively short spans on a geological timescale but are far too long to be used for any revealing calculations of the likely risks per year (to say nothing about per hour of exposure). Moreover, global consequences would be very different if such an object struck the Pacific Ocean near Antarctica than if it hit Western Europe or eastern China. In the first case, much of the damage would come from a monstrous tsunami but (depending on the asteroid's size) there might be little dust entering the atmosphere. In the second and third cases the impact would instantly obliterate large concentrations of the population and industrial activity, and throw enormous masses of powdered rocks into the atmosphere, creating a pronounced planet-wide cooling.

Americans should not worry either about supernovas or asteroids, but if they want to scare themselves by thinking about an inevitable natural catastrophe (and one that would emanate from one of the country's cherished places!) then they should consider another mega-eruption of the Yellowstone supervolcano.[72] Geological evidence shows nine eruptions during the past 15 million years, with the last three known eruptions occurring 2.1 million, 1.3 million, and 640,000 years ago. Of course, the dating of just three events offers no basis for predicting any periodicity, but still a thought intrudes: taking the mean interval of 730,000 between eruptions we would have still 90,000 years of waiting left, but if the first interval was 800,000 years and the second one was 660,000 years then a similar shortening would indicate the next span being some 520,000 years—and a new eruption would be already more than 100,000 years overdue!

And whatever the interval, the consequences would depend on the magnitude of the eruption, on its duration, and on prevailing winds. The last eruption released about 1,000 cubic kilometers of volcanic ash, and the prevailing northwesterly winds would carry the plume over Wyoming (where the deepest deposits could be several meters thick), Utah, and Colorado and onto the Great Plains, affecting states from South Dakota to Texas and burying some of the country's most productive agricultural land under 10–50 centimeters of ash. The combination of advance warning (due to constant seismic monitoring) and a weaker, prolonged eruption might make large-scale evacuation possible, and the loss of housing, infrastructure, and cultivable land would then be far greater than any immediate casualties. A thin covering of volcanic ash could be plowed into the soil (and actually improve its fertility), but thicker layers would be unmanageable and they would pose additional dangers once dislodged by rains and snow melt, resulting in silting and flooding and creating problems for decades to come.

Perhaps the best example of a natural risk that would not directly kill anybody, but that would cause enormous planet-wide disruptions resulting in a large number of indirect casualties, is the possibility of a catastrophic geomagnetic storm caused by a coronal mass ejection.[73] The corona is the outermost layer of the Sun's atmosphere (it can be seen without special instruments only during a total solar eclipse) and is, paradoxically, hundreds of times hotter than the Sun's surface. Coronal mass ejections are enormous (billions of tons) expulsions of explosively accelerated material that carry an embedded magnetic field whose strength greatly surpasses that of background solar wind and the interplanetary magnetic field. Coronal ejections begin with the twisting and reconfiguration of the magnetic field in the layer's lower part; they produce solar flares and can travel (expanding as they advance) at speeds as slow as less than 250 km/s (arriving at the Earth in nearly seven days) and as fast as almost 3,000 km/s (reaching the Earth in as little as 15 hours).

The largest known coronal mass ejection began on the morning of September 1, 1859, while Richard Carrington, a British astronomer, was observing and drawing a large solar sunspot that emitted a sizable, kidney-shaped white flare.[74] That was nearly two decades before the first telephones (1877) and more than two decades before the first centralized commercial generation of electricity (1882), and hence the notable effects were only intense auroras and disruptions of the newly expanding telegraph network whose laying began in the 1840s: wires were sparking, messaging was interrupted or continued in bizarrely truncated ways, operators got electric shocks, some fires started accidentally.

Some of the subsequent strongest events took place on October 31–November 1, 1903 and May 13–15, 1921, when the extent of both wired telephone links and electricity grids was still fairly limited even in Europe and North America, and very sparse elsewhere. But we got a preview of what a substantial coronal mass ejection could do today in March 1989 when a much smaller (a non-Carrington) event knocked out Quebec's entire power grid, serving 6 million people, for nine hours.[75] More than three decades later we have become much more vulnerable: just think of everything electronic, from mobile phones to e-mail to international banking, and about GPS-guided navigation on every vessel and airplane and now also on tens of millions of cars.

We would find out before it hits: our constant surveyance of the Sun's activity would instantly detect any mass ejection and provide at least 12–15 hours of prestrike warning. But only when the ejection reaches the point where we have stationed the Solar and Heliospheric Observatory (SOHO), about 1.5 million kilometers away from the Earth, could we gauge its intensity; and by then the time to react would be reduced to less than an hour, perhaps even to just 15 minutes.[76] Even limited damage would mean hours or days of disrupted communications and grid operations, and a massive geomagnetic storm would sever all of these links on a global scale, leaving us without electricity, without information, without transportation, without the ability to make credit card payments or to withdraw money from banks.

What would we do if the complete restoration of all these vital but severely crippled infrastructures took years, perhaps even a decade, to accomplish? Estimates of global damage differ by an order of magnitude, from $2 trillion to $20 trillion[77]—but that refers only to expenditures, not to the value of lives lost during prolonged spells without communication, lights, air conditioning, hospital equipment, refrigeration, and industrial production (and hence also without adequate inputs into crop cultivation).

There is some good news. A 2012 study estimated a 12 percent probability of another Carrington Event during the coming ten years—or a one-in-eight chance, and it emphasized that the rarity of these extreme events makes their rate occurrence difficult to estimate "and prediction of a specific future event is virtually impossible."[78] Given this uncertainty, it is not surprising that in 2019 a group of scientists in Barcelona calculated the risk to be no greater than 0.46–1.88 percent during the 2020s, and hence even the highest rate would mean odds of 1 in 53, a considerably more comforting probability.[79] And in 2020 a Carnegie Mellon group offered an even lower estimate, putting a decadal (10-year) probability of between 1 percent and 9 percent for an event of at least the size of the large 2012 event, and between 0.02 percent and 1.6 percent for the size of the 1859 Carrington Event.[80] While many experts are well aware of these odds and of the enormity of the potential consequences, this is clearly one of those risks (much like a pandemic) for which we cannot ever be adequately prepared: we just have to hope that the next massive coronal ejection event will not equal or surpass the Carrington Event.

While this may not be what the world wants to hear at this time, it is an unfortunate truth that viral pandemics are guaranteed to reappear with relatively high frequency and, although sharing inevitable commonalities, they are unpredictably specific in their impacts. In early 2020 the world had about a billion people older than 62 years, and they had all lived through three viral pandemics in a single lifetime: 1957–1959 (H_2N_2), 1968–1970 (H_3N_2), and 2009 (H_1N_1).[81] The best reconstruction of the total mortality for the 1957–1959 pandemic was 38/100,000 (1.1 million deaths; global population 2.87 billion), the 1968–1970 pandemic had a mortality of 28/100,000 (1 million deaths; global population 3.55 billion), while the 2009 event had low virulence and mortality no higher than 3/100,000 (about 200,000 deaths; global population 6.87 billion).[82]

Arrival of the next event was just a matter of time, but as already noted we are never prepared for these (relatively) low-frequency threats. The World Economic Forum's ranking of top global risks, prepared annually between 2007 and 2015, led with asset price collapse, financial crisis, and major systemic financial failure eight times (obvious echoes of 2008) and water crises once, while pandemic threat did not appear among the top three risks even once.[83] So much for the collective foresight of global decision-makers! And when COVID-19 (caused by SARS-CoV-2) arrived, the World Health Organization waited until March 11, 2020 to proclaim a global pandemic, and its early advice (echoed by many governments) was against suspending international flights and against wearing masks.[84]

Obviously, we will be able to quantify the total COVID-19 mortality only after this latest pandemic ends. Meanwhile, the best way to assess the recurrent pandemic burden is to compare it to global seasonal influenza-associated respiratory mortality. The most detailed assessment for the years 2002–2011 found a mean of 389,000 deaths (ranging between 294,000 and 518,000) after excluding the 2009 pandemic season.[85] This means that seasonal influenza accounts for about 2 percent of all annual respiratory deaths, and that its mortality rate averages 6/100,000—or 15–20 percent of the death rates recorded in the two late 20th-century pandemics (1957–1959, 1968–1970). Inversely stated, the first pandemic exacted a more than six times higher and the second one a nearly five times higher relative death toll than seasonal influenza.

Moreover, there is an important difference in age-specific mortality. Seasonal flu mortality is, almost without exception, highly skewed toward old age, with 67 percent of all deaths among people over 65. In contrast, the infamous second wave of the 1918 pandemic disproportionately targeted people in their 30s; the 1957–1959 pandemic had a

U-shaped mortality frequency, disproportionately affecting ages 0–4 and 60+; while COVID-19 mortality has been, much like seasonal influenza, highly concentrated in the 65+ cohort, especially among those with significant comorbidities, and it has left children remarkably unaffected.[86]

We know that many excess deaths among older people cannot be prevented: that is a part of the price we must pay for our very successful efforts to extend life expectancy (in many affluent countries by more than 15 years since the 1950s).[87] A death certificate may say COVID-19 or viral pneumonia, but that is just the proximate label—the real cause is that most of us have not been designed to be without underlying health problems as we keep pushing the limits of life expectancy. Provisional COVID-19 data from the CDC make that clear: during the week of the peak US COVID-19 mortality (ending on April 18, 2020), people over 65 years of age accounted for 81 percent of all deaths, and those younger than 35 years for a mere 0.1 percent.[88] This situation is quite different from the 1918–1920 pandemic when as many as 50 million people died. We now know that most of those deaths were due to bacterial pneumonia: some 80 percent of cultures taken from preserved lung tissue samples contained bacteria causing secondary lung infection—and at that time, nearly a quarter-century before the availability of antibiotics, we had no treatment for that condition.[89]

Moreover, people with tuberculosis were more likely than others to die of influenza, and this link also helps to explain the 1918–1920 pandemic's unusual middle-age mortality as well as its clear maleness (due to the differential incidence of tuberculosis).[90] Because tuberculosis has been essentially eradicated in all affluent countries and because pneumonia is treatable with antibiotics, we can avoid the repetition of high mortalities, but even with our annual influenza vaccination drives we cannot prevent significant seasonal mortality, and the survival of the oldest cohorts will be challenged every time there is a global pandemic. This is a largely self-inflicted risk, the obverse of enjoying longer life expectancy, and we can minimize it by isolating the most vulnerable individuals and by developing better vaccines—but we cannot eliminate it.

Some lasting attitudes

Where risk is concerned, many truisms seem to be permanent. As individuals, we can exercise some control. Many people do not find it difficult to abstain from smoking, consuming alcohol and drugs, and prefer to stay home rather than sharing a cruise ship with 5,000 passengers and 3,000 crew in the midst of a coronavirus or norovirus outbreak. Others crave all of the above, and it is astonishing how many people do not reduce even the most easily—and inexpensively—reducible risks. Always wearing a seat belt, never speeding, being a defensive driver, and installing smoke, carbon monoxide, and natural gas sensors in dwellings are no-cost or very low-cost ways to reduce the risks of driving and of living in structures heated by the combustion of fossil fuels.

Moreover, most people and most governments find it difficult to deal properly with low-probability but high-impact (high-loss) events. Buying basic house insurance is one thing (often it is mandatory); investing in earthquake-resistant structures—be it as individuals or as societies—to minimize the impact of what is likely to be a once-in-a-century event is quite another matter. California has a subsidized seismic retrofit program for pre-1980 houses (bolting, or bolting and bracing, the house to its foundation in compliance with 2016 building code)—but most jurisdictions that face similar seismic risks do not.[91]

But it is difficult, if not impossible, to avoid many exposures, because (as already noted) in some cases there is no clear dichotomy between voluntary and involuntary risks. And most risks are beyond our control. We cannot choose our parents and hence avoid a genetic predisposition to a large number of common and rare diseases, including some cancers, diabetes, cardiovascular problems, asthma, and several autosomal recessive disorders including cystic fibrosis, sickle cell anemia, and Tay-Sachs disease.[92] In order to greatly reduce the risks of all local or regional natural disasters, we would have to eliminate large areas of the planet—above all, those subject to recurrent mega-earthquakes and volcanic eruptions (the Pacific Ring of Fire), destructive cyclonic winds, and extensive flooding—as places of human habitation.[93]

Because that is clearly impossible on an increasingly crowded planet, the only way to improve the odds of survival under those conditions is to take precautions—earthquake-proof (steel-reinforced) buildings will not bury people as surrounding structures collapse; tornado shelters will save families so they can rebuild their leveled homes—and put in place effective early-warning systems and mass-scale evacuation plans to reduce the loss of life caused by cyclones, floods, and volcanic eruptions. While these measures could potentially save not just hundreds but hundreds of thousands of lives, we have limited defenses or are entirely powerless against many large-scale catastrophes ranging from massive earthquake-triggered tsunami to mega-volcanic eruptions, and from prolonged regional droughts to the Earth's encounters with asteroids or comets.

Another set of truisms applies to our risk assessment. We habitually underestimate voluntary, familiar risks while we repeatedly exaggerate involuntary, unfamiliar exposures. We constantly overestimate the risks stemming from recent shocking experiences and underestimate the risk of events once they recede in our collective and institutional memory.[94] As I already noted, about a billion people have lived through three pandemics, but when COVID-19 struck references were made overwhelmingly to the 1918 episode, as the three more recent (but less deadly) pandemics—unlike the widely remembered fear of polio during the 1950s or AIDS in the 1980s—have left no or only the most superficial impressions.[95]

There are obvious explanations for this amnesia. The 2009 pandemic was essentially undistinguishable from a seasonal influenza, and neither in 1957–1959 nor in 1968–1970 did we resort to near-complete national or continental lockdowns. Inflation-adjusted statistics of global and US economic product show no drastic reversal of long-term growth rates during either of the two late 20th-century pandemics.[96] Moreover, the latter episode coincided with a significant expansion of international air travel: the first wide-body jetliner, the Boeing 747, flew for the first time in 1969.[97] And, perhaps most importantly, we had no 24/7 cable TV news with its morbid attachment to announcing running death counts, no internet rife with ridiculous claims about causes and cures and with conspiracy theories, and hence no ahistorical but hysterical ways of modern news diffusion.

As COVID-19 demonstrated yet again (and on scales that must have surprised even those who do not expect any good news), we are repeatedly caught ill-prepared for dealing with recurrent high-impact but relatively low-frequency risks such as viral pandemics that take place once in a decade, once in a generation, or once in a century. How would we then cope (all reports and analyses aside) with another Carrington Event, or with an asteroid hitting the ocean near the Azores and causing a massive circum-Atlantic tsunami of the same magnitude as the one caused by the 2011 Tōhoku earthquake—that is, up to 40 meters high and traveling up to 10 kilometers inland?[98]

And the lessons we derive in the aftermath of major catastrophic events are decidedly not rational. We exaggerate the probability of their recurrence, and we resent any reminders that (setting the shock aside) their actual human and economic impact has been comparable to the consequences of many risks whose cumulative toll does not raise any extraordinary concerns. As a result, fear of another spectacular terrorist attack led the US to take extraordinary steps to prevent it. These included multitrillion-dollar wars in Afghanistan and Iraq, fulfilling Osama bin Laden's wish to draw the country into stunningly asymmetrical conflicts that would erode its strength in the long run.[99]

Public reaction to risks is guided more by a dread of what is unfamiliar, unknown, or poorly understood than by any comparative appraisal of actual consequences. When these strong emotional reactions are involved, people focus excessively on the possibility of a dreaded outcome (death by a terrorist attack or by a viral pandemic) rather than trying to keep in mind the probability of such an outcome taking place.[100] Terrorists have always exploited this reality, forcing governments to take extraordinarily costly steps to prevent further attacks while repeatedly neglecting to take measures that could have saved more lives at a much lower cost per averted fatality.

There is no better illustration of neglected low-cost measures to save lives than the American attitude to gun violence: not even the most shocking iterations of familiar, all-too-well-known mass murders (I always think first of the 26 people, including 20 six- and seven-year-old children, shot in 2012 in Newtown, Connecticut) have been able to change the laws, and during the second decade of the 21st century about 125,000 Americans were killed by guns (the total for homicides, excluding suicides): that is the equivalent of the population of Topeka, Kansas or Athens, Georgia or Simi Valley, California—or of Göttingen in Germany.[101] In contrast, 170 Americans died in all terrorist attacks in the US during the second decade of the 21st century, a difference of nearly three orders of magnitude.[102] When we compare this to motor vehicle accidents, the toll is even more unevenly distributed: as we saw earlier, compared to Asian American females, Native American men are about five times more likely to die in their cars, but African American males are about 30 times more likely to be killed by firearms.[103]

Do I have a helpful parting insight? Perhaps, as long as we recognize these fundamental realities: asking for a risk-free existence is to ask for something quite impossible—while the quest for minimizing risks remains the leading motivation of human progress.

6. Understanding the Environment

The Only Biosphere We Have

This chapter's subtitle is deliberately preventive. I refuse to consider any near-term possibility of leaving the Earth and setting up a civilization on another planet. I do this because, in this post-factual world, musings about soon finding a new celestial abode—most notably, the terraforming of Mars[1]—have been presented as possible options to deal decisively with the problems of the third planet orbiting the Sun. This is yet another favorite topic of the sci-fi genre that will remain confined to its stories: even if we had inexpensive means of interplanetary transport and somehow mastered the construction of Martian bases, we could not create a suitable atmosphere—the processing of Martian polar caps, minerals, and soil would yield only about 7 percent of all the CO_2 that would be needed in order to warm the planet and make its prolonged colonization possible.[2]

Of course, the true believers can call on another sci-fi trick that could enable the colonization of Mars: creating radically genetically re-engineered humans, new super-organisms endowed with qualities of terrestrial tardigrades, tiny eight-legged invertebrates living on grass and in wet ditches. Such organisms would be able to cope not only with the thin atmosphere (its pressure is less than 1 percent of the terrestrial value) but also with the high radiation received by the poorly shielded red planet.[3]

Returning to the real world, if our species is to survive, never mind to flourish, for at least as long as high civilizations have been around (that is, for another 5,000 or so years), then we will have to make sure that our continuing interventions do not imperil the long-term habitability of the planet—or, as modern parlance has it, that we do not transgress safe planetary boundaries.[4]

The list of these critical biospheric boundaries includes nine categories: climate change (now interchangeably, albeit inaccurately, called simply global warming), ocean acidification (endangering marine organisms that build structures of calcium carbonate), depletion of stratospheric ozone (shielding the Earth from excessive ultraviolet radiation and threatened by releases of chlorofluorocarbons), atmospheric aerosols (pollutants reducing visibility and causing lung impairment), interference in nitrogen and phosphorus cycles (above all, the release of these nutrients into fresh and coastal waters), freshwater use (excessive withdrawals of underground, stream, and lake waters), land use changes (due to deforestation, farming, and urban and industrial expansion), biodiversity loss, and various forms of chemical pollution.

Providing systematic reviews of all of these concerns—and setting them in their appropriate historical and environmental perspectives—is a task for a major book, not for a single chapter (unless it consisted of superficial summaries). Instead, I have decided to give this chapter a decidedly utilitarian tilt and focus on just a few key existential parameters, starting with the environmental circumstances of three irreplaceable existential requirements—breathing, drinking, and eating. Provision of these three preconditions of our existence depends on natural goods and services: on the oxygenated atmosphere and its incessant circulation; on water and its global cycle; and on soils, photosynthesis, biodiversity, and flows of plant nutrients. In turn, their provision affects natural goods and services.

As we will see, these effects range from marginal (oxygen concentrations in the atmosphere are in no danger because of fossil fuel combustion) to obviously negative (excessive water extraction from ancient deep aquifers; serious water pollution generated by food production, cities, and industries) to outright destructive (overgrazing in arid regions leading to desertification; new cropland displacing tropical forests or grasslands).

Oxygen is in no danger

Breathing is the regular delivery of oxygen, carried from our lungs by hemoglobin to all cells in the body to energize our metabolism. No provision of a natural resource is as critical for our survival:

duration of bearable voluntary apnea (the cessation of breathing) varies, but if you have never trained yourself to prolong your breathless periods you will find out that you can last as little as 30 seconds and typically no longer than about a minute or so. You might have read about freediving, where men and women risk their lives by holding their breath and diving, without any breathing apparatus, as deep as they can endure (with or without fins), or about static apnea competitions where the contestants lie motionless in a pool of water and hold their breath. The latter record for men is nearly 12 minutes, for women 9 minutes, while hyperventilating with pure oxygen for up to half an hour before an attempt doubles apneic time to more than 24 minutes for men and 18½ minutes for women.[5]

This passes for sport in the 21st century despite the fact that brain cells start to die within five minutes of cerebral hypoxia and that an only slightly longer period can cause severe damage or death. Oxygen, after all, is the most acutely limiting resource for human survival. Our species, like all other chemoheterotrophs (organisms that cannot internally produce their own nutrition), requires its constant supply. The resting frequency of breathing is 12–20 inhalations a minute, and the daily adult per capita intake averages almost 1 kilogram of O_2.[6] For the global population, that translates to an annual intake of about 2.7 billion tons of oxygen a year, an utterly insignificant fraction (0.00023 percent) of the element's atmospheric presence of about 1.2 quadrillion tons of O_2—and the exhaled CO_2 is readily used by photosynthesizing plants.

The beginnings of the oxygenated atmosphere go back to what has become known as the Great Oxidation Event, which began about 2.5 billion years ago.[7] During that period, oxygen released by oceanic cyanobacteria began to accumulate in the atmosphere, but it took a long time before the gases reached their modern concentrations. During the past 500 million years, atmospheric oxygen levels fluctuated widely, being as low as about 15 percent and as high as 35 percent before they declined to today's nearly 21 percent of the Earth's atmosphere by volume.[8] As well as there being absolutely no danger of people or animals appreciably reducing this level through breathing, there is also no danger of too much oxygen being consumed by even the greatest conceivable burning (rapid oxidation) of the Earth's plants.

The Earth's terrestrial plant mass contains on the order of 500 billion tons of carbon and even if all of it (all forests, grasslands, and crops) were burned at once, such a mega-conflagration would consume only about 0.1 percent of the atmosphere's oxygen.[9] And yet, during the summer of 2019, when large areas of the Amazon rainforest were burning, news media and politicians tried to frighten the scientifically illiterate masses into believing that the world would start suffocating. One among many, on August 22, 2019 the French president Emmanuel Macron tweeted:[10]

> Our house is burning. Literally. The Amazon rain forest—the lungs which produces 20% of our planet's oxygen—is on fire. It is an international crisis. Members of the G7 Summit, let's discuss this emergency first order in two days!

There was no emergency G7 Summit in two days (or even two months: a good thing too, as if that could fix anything!) and the world keeps on breathing. Depending on where you happen to be on this specific judgmental scale, deliberate burning of the Amazon rainforest is either a highly regrettable and completely misguided policy or an unpardonable crime against the biosphere—but know that it is not an act that will deprive the planet of its oxygen.

This misinformation also illustrates a far wider problem—namely why are we not relying on well-established scientific facts, and instead, why do we let assorted tweets drive public opinion? Appraisals of the environment are perhaps even more prone to unwarranted generalization, biased interpretation, and outright misinformation than those of energy and food production. This tendency must be condemned and resisted: we will not succeed if our actions are based on myths and misinformation. Admittedly, the underlying science is often complex and many verdicts are uncertain and resolute judgments are unadvisable—but not in this particular case.

Most obviously, lungs do not produce oxygen, they process it: the function of lungs is to enable gas exchange as atmospheric O_2 enters the bloodstream and CO_2, the most voluminous gaseous product of metabolism, leaves it. In the process, lungs (as any other organ) must consume oxygen, but it is not easy to measure how much of it they need—that is, to separate their requirement from the overall intake. The best way to find out is during a total cardiopulmonary bypass, when lung circulation is temporarily separated from systemic blood flow: this shows that lungs consume about 5 percent of the total oxygen we inhale.[11] And while Amazonian trees, as any terrestrial plants, produce O_2 during diurnal photosynthesis, they—again, much as any other photosynthesizing organism—consume virtually all of this oxygen during nocturnal respiration, the process that uses photosynthate to produce energy and compounds for plant growth.[12]

Every year, at least 300 billion tons of oxygen are absorbed and a similar amount is released by terrestrial and marine photosynthesis.[13] These flows, as well as much smaller flows resulting from the burial and oxidation of organic matter, are not perfectly balanced on a daily or seasonal basis, but over the long run they cannot be too far off, otherwise we would have substantial net gains or losses of the element. Instead, oxygen's atmospheric presence has been remarkably stable. Images of burning Amazonian forest, Australian scrubland, Californian hillsides, or Siberian taiga are not ominous harbingers of an atmosphere deprived of the gas we need to inhale at least a dozen times a minute.[14] Massive forest fires are destructive and harmful in many ways, but they are not going to suffocate us because of a lack of oxygen.

Will we have enough water and food?

In contrast, the provision of the second-most acutely required natural input should be very high on our list of environmental worries—and not because there is any absolute shortage of this critical resource but because it is unevenly distributed and because we have not managed it well. And that is an understatement—we waste water enormously and, so far, we have been slow to adopt many effective changes that would reverse undesirable habits and trends. As we will see, water supply is thus a perfect example of an almost universally mismanaged resource, with the added complication of highly uneven access.[15]

At least we do not have to drink as often as we breathe, a dozen times a minute, not even a dozen times a day—but the provision of adequate volumes of drinking water (which, depending on gender, age, body size, ambient temperature, and excluding extreme activities, is mostly between 1.5 and 3 liters a day) is a matter of basic survival.[16] No rehydration for a day is a trying experience, for two days it becomes perilous, for three days it is usually fatal. Beyond this existential necessity, translating to a per capita average of about 750 kilograms (or liters, or 0.75 cubic meters) of water a year, there are several other—and much more voluminous—water needs: for personal hygiene, cooking, and laundry (even without an indoor toilet, these categories add up to the minima of 15–20 liters a day, or about 7 cubic meters a year), for productive activities, and, above all, for growing food.[17]

Different sectors of water use (agriculture, thermal electricity generation, heavy industries, light manufacturing, services, household) and different categories of water complicate intranational and international comparisons. Blue water includes rainfall entering rivers, water bodies, and groundwater storage that gets incorporated into products or evaporates; the green water footprint accounts for water from precipitation that is stored in soil and subsequently evaporated, transpired, or incorporated by plants; grey water includes all the freshwater required to dilute pollutants in order to meet specific water-quality standards.

This is why national per capita consumption is the best (the most exhaustive) way to assess water footprints: it adds green, blue, and grey water components as well as all virtual water (water that was required for the growth or production of imported food and manufactured goods).[18] Domestic blue water use (all values are in cubic meters per year per capita) ranges from just over 29 in Canada and 23 in the USA to about 11 in France, 7 in Germany, and about 5 in China and India, and to less than 1 in many African countries.[19] The total water footprint of national consumption reflects specific shares of water used in agriculture (obviously highest in countries with extensive irrigation) and industrial production. As a result, economies with very different climates and sectoral consumption—Canada and Italy, Israel and Hungary—have similar consumption totals (in all of these cases, between 2,300 and 2,400 cubic meters/year/capita). Food imports incorporate considerable amounts of green water, and hence the two countries with the highest dependence on imported food—Japan, South Korea—are also the highest users of virtual water.

Not surprisingly, water's critical role in national economies in general, and in food production in particular, has resulted in many comprehensive appraisals of its availability, sufficiency, scarcity, and vulnerability. At the beginning of the 21st century, water-stressed populations totaled as low as 1.2 billion and as high as 4.3 billion—that is, between 20 and 70 percent of all humanity.[20] Similarly, during the second decade of the 21st century two different measures of water scarcity indicated that the affected populations were between 1.6 and 2.4 billion people.[21] Given these major differences in present assessments, it is impossible to offer strongly defensible conclusions about the future.

There are also many uncertainties concerning the future provision of food. No other human activity has transformed the Earth's ecosystems to a greater extent than the production of food. It already adds up to about a third of the planet's non-glaciated land, and further impacts are inevitable.[22] The combined area devoted to food production is now more than twice as large as it was a century ago, but in all affluent economies the land under cultivation has either stabilized or has slightly decreased, while the overall global growth of new farmland has slowed down considerably.[23] Given the continent's still-high fertility rates, further expansion of cultivated land will be inevitable in Africa, but only limited extensions should take place in most of Asia, while Europe, North America, and Australia (with already-excessive food production and aging populations) should see further declines in cultivated land.

The amount of land used in food production could be reduced with the combination of better farming practices, reduced food waste, and the widespread adoption of moderate meat consumption. As already explained in chapter 2, reversion to preindustrial farming is inconceivable in a world of nearly 8 billion people, but getting higher yields with existing inputs (agricultural intensification) conforms to a long-established trend, and the elimination of many wasteful practices could produce higher yields even with reduced fertilizer or pesticide use. A convincing large-scale, decade-long (2005–2015) demonstration of this included nearly 21 million farmers cultivating about a third of China's cropland: they were able to raise staple grain yields by 11 percent while reducing the application of nitrogen per hectare by 15–18 percent.[24]

If land is not a limiting resource, and if we have the know-how to manage water supply, what are the prospects for providing the macronutrients our crops need while also limiting the environmental impact of applying nitrogen and phosphorus? As already explained, the Haber-Bosch synthesis of ammonia made it possible to provide a reactive form of nitrogen, the leading macronutrient, in any desirable quantity.[25] We can also provide adequate amounts of the two mineral macronutrients, potassium and phosphorus. The US Geological Survey puts potassium resources at about 7 billion tons of K_2O (potassium oxide) equivalent; reserves are about half that amount, and at the current rate of production these reserves would last for nearly 90 years.[26]

During the last 50 years there have been periodic comments about imminent phosphorus shortages, some even raising the inevitability of starvation in a matter of decades.[27] Concerns about wasting a finite resource are always appropriate, but there is no imminent phosphorus crisis. According to the International Fertilizer Development Center, the world's phosphate rock reserves and resources are adequate to meet fertilizer demand for the next 300–400 years.[28] The US Geological Survey puts the world resources of phosphate rock at more than 300 billion tons, sufficient for more than 1,000 years at the current rate of extraction.[29] And the International Fertilizer Industry Association "does not believe that peak phosphorus is a pressing issue, or that phosphate rock depletion is imminent."[30]

The real concern about plant nutrients is the environmental (and hence economic) consequences of their unwanted presence in the environment, mostly in water. Phosphorus from fertilizers is lost through soil erosion and precipitation runoff and it is released in waste produced by domestic animals and people.[31] Because water (whether fresh or ocean) normally has very low concentrations of this element, its additions lead to eutrophication, the enrichment of waters with previously scarce nutrients that results in the excessive growth of algae.[32] Losses of nitrogen from fertilized farmland (and from animal and human waste) also cause eutrophication, but aquatic photosynthesis is more responsive to phosphorus additions. Neither primary sewage treatment (sedimentation removes 5–10 percent of phosphorus) nor secondary removal (filtration captures 10–20 percent) prevents eutrophication, but phosphorus can be removed by using coagulating agents or by microbial processes, then turned into crystals and reused as fertilizer.[33]

As already explained, the worldwide efficiency of nitrogen uptake by crops has declined to less than 50 percent, and to below 40 percent in China and France. In conjunction with phosphorus, soluble nitrogen compounds contaminate waters and support excessive algal growth. Decomposing algae consume oxygen dissolved in seawater and create oxygen-less (anoxic) waters where fish and crustaceans cannot survive. These oxygen-depleted zones are prominent along the eastern and southern coasts of the United States and along coasts in Europe, China, and Japan.[34] There are no easy, inexpensive, and rapid solutions to these environmental impacts. Better agronomic management (crop rotations, split applications of fertilizers to minimize their losses) is essential, and reduced meat consumption would be the single-most important adjustment as it would lower the need for producing feed grains—but sub-Saharan Africa will need much more nitrogen and phosphorus if it is to avoid chronic dependence on food imports.

And any longer-term assessment of the three existential necessities—atmospheric oxygen, water availability, and food production—must consider how their provision could be affected by the unfolding process of climate change, a gradual transformation that will leave its mark on the biosphere in myriad ways: the impacts go far beyond higher temperatures and rising ocean levels, the two changes that are most often referenced by the media. I will not revisit a long list of predicted impacts, from broiling cities to rising oceans, from desiccated crops to melted glaciers. That has been done, in measured ways as well as in hysterical fashion, too many times.

Instead, I will take a utilitarian—and unorthodox—approach. I will start by explaining the life-enabling necessity of the greenhouse gas effect, without which the Earth's surface would be permanently frozen and which we have, unintentionally, enhanced by a combination of actions—with the combustion of fossil fuels being the most important driver of anthropogenic global warming. Then I will explain how, contrary to a common perception, modern science identified this phenomenon more than a century ago, how we have ignored clearly stated potential risks for generations, how we have been, so far, unwilling to commit to any effective action to change the course of global warming—and how extraordinarily challenging such a shift would be.

Why the Earth is not permanently frozen

As we saw in the first chapter, the abundance of fossil fuels and their increasingly more efficient conversions have been the dominant energizers of modern economic growth, bringing us the benefits of greater longevity and richer lives—but also concerns about the long-term effects of CO_2 emissions on the global climate (commonly referred to as global warming). Simple physics explains our worries about the environmental consequences of planetary warming. We are concerned about too much of something without which we could not be alive: the greenhouse effect. This existential imperative is the regulation of the Earth's atmospheric temperature by a few trace gases—above all by carbon dioxide (CO_2) and methane (CH_4). Compared to the two gases that make the bulk of the atmosphere (nitrogen at 78 percent, oxygen at 21 percent), their presence is negligible (small fractions of a percent) but their effect makes the difference between a lifeless, frozen planet and a blue and green Earth.[35]

The Earth's atmosphere absorbs the incoming (short-wave) solar radiation and radiates (longer waves) to space. Without it, the temperature of the Earth would be -18°C, and hence our planet's surface would be perpetually frozen. Trace gases change the planet's radiation balance by absorbing some of the outgoing (infrared) radiation and raising the surface temperature. This allows for the existence of liquid water, whose evaporation puts water vapor (another gas that absorbs outgoing infrared, invisible waves) into the atmosphere. The overall result is that the Earth's surface temperature is 33°C higher than it would be in the absence of trace gases and water vapor, and the global mean temperature of 15°C supports life in its many forms.

Labeling this natural phenomenon as the "greenhouse effect" is a misleading analogy, because the heat inside a greenhouse is there not only because the glass enclosure prevents the escape of some infrared radiation but also because it cuts off air circulation. In contrast, the natural "greenhouse effect" is caused solely by the interception of a small share of outgoing infrared radiation by trace gases—while the global atmosphere remains in constant, unimpeded, and often violent motion. Water vapor is by far the most important absorber of outgoing radiation, and hence it is the gas that has been responsible for most of the atmospheric warming in the past and it will remain so in the future. Water vapor is the principal generator of the natural greenhouse effect—but water vapor is not the cause of atmospheric warming because it does not control atmospheric temperature. In fact, it is the other way around: the changing temperature determines how much water can be present as a gas (the humidity of air increases with rising temperatures) and how much it condenses to liquid (condensation increases as the temperature drops).

The Earth's natural warming is controlled by trace gases whose concentration is not affected by the ambient temperature—that is, they do not condense and precipitate as temperatures decline. But the relatively small warming they cause increases evaporation and raises atmospheric water concentrations, and this feedback results in additional warming. The natural trace gas effect has always been dominated by carbon dioxide (CO_2), with smaller contributions made by methane (CH_4), nitrous oxide (N_2O), and ozone (O_3)—the latter known to many via the eponymous ozone layer. Human actions began to affect the concentrations of several trace gases—creating an additional, man-made (anthropogenic) greenhouse gas effect—thousands of years ago, as soon as settled societies adopted farming and began using wood (and the charcoal made from it) in households and when smelting metals and making bricks and tiles. The conversion of forests to cropland released additional CO_2, and cultivation of rice in flooded fields produced additional CH_4.[36]

But the impact of these anthropogenic emissions became significant only with the increasing pace of industrialization. Rising CO_2 emissions, which cause an accelerated anthropogenic greenhouse gas effect, have been driven primarily by the combustion of fossil fuels and by the production of cement. Methane emissions (from rice fields, landfills, cattle, and natural gas production) and nitrous oxide (originating mostly from the rising application of nitrogenous fertilizers) are the other notable anthropogenic sources of greenhouse gases. Reconstructions of their past atmospheric concentrations show the sudden increase brought on by industrialization.

For centuries before 1800, CO_2 levels fluctuated narrowly at close to 270 parts per million (ppm)—that is, 0.027 percent by volume. By 1900 they rose slightly to 290 ppm, a century later they were nearly 375 ppm, and in the summer of 2020 they rose above 420 ppm, more than a 50 percent increase above the late 18th-century level.[37] Preindustrial methane levels were three orders of magnitude lower—less than 800 parts per billion (ppb)—but they have more than doubled, to nearly 1,900 ppb by 2020, while nitrous oxide concentrations rose from about 270 ppb to more than 300 ppb.[38] These gases absorb the outgoing radiation to different degrees: when their impacts are compared over a 100-year period, releasing a unit of CH_4 has the same effect as releasing 28–36 units of CO_2; for N_2O, the multiplier is between 265 and 298. A handful of new man-made industrial gases—above all chlorofluorocarbons (CFCs, in the past used in refrigeration) and SF_6 (an excellent insulator used in electrical equipment)—exert a far stronger effect, but fortunately they are present only in minuscule concentrations and the production of CFCs was gradually outlawed by the 1987 Montreal Protocol.[39]

CO_2 (mostly emitted from fossil fuel combustion, with deforestation being another major source) accounts for about 75 percent of the anthropogenic warming effect, CH_4 for about 15 percent, and the rest is mostly N_2O.[40] The continued rise of greenhouse gas emissions will eventually lead to temperatures high enough to cause many negative environmental impacts, engendering considerable social and economic costs. Contrary to a widely held impression, this is not a recent conclusion arising from a better understanding provided by complex climate change models executed by supercomputers. We knew this not only long before the first models of global atmospheric circulation (the precursors of all global warming simulations) were introduced during the late 1960s, but generations before the first electronic computers were even built.

Who discovered global warming?

If you check Google's Ngram Viewer for the appearance of "global warming," you will discover the virtual absence of the phrase before 1980 followed by a steep rise in frequency, quadrupling in the two years before 1990. The "discovery" of carbon dioxide–induced global warming by the media, public, and politicians came in 1988, spurred by America's warm summer and by the establishment of the Intergovernmental Panel on Climate Change (IPCC), by the UN's Environment Programme (UNEP), and by the World Meteorological Organization (WMO). This led to a still-rising tide of scientific papers, books, conferences, think tank studies, and reports prepared by governments and international organizations, including the IPCC's periodic state-of-the-art reviews.

By 2020, a Google search returned more than a billion items for "global warming" and "global climate change"—this frequency is an order of magnitude higher than those for such recently fashionable news items as "globalization" or "economic inequality," or for such existential challenges as "poverty" and "malnutrition." Moreover, almost since the very beginning of the media's interest in this complex process, the coverage of global warming has been replete with poorly communicated facts, dubious interpretations, and dire predictions, and over time it has definitely acquired a distinctly more hysterical, even outright apocalyptic, flavor.

Uninformed observers would have to conclude that these warnings of an unfolding global catastrophe reflect the very latest scientific findings based on a combination of previously unavailable satellite observations and on the forecasts using complex global climate models whose execution has been made possible by the rise of computing power. But while our latest monitoring and modeling are certainly more advanced, there is nothing new either about our understanding of the greenhouse effect or about the consequences of steadily increasing emissions of greenhouse gases: in principle, we have been aware of them for more than 150 years, and in a clear and explicit manner for more than a century!

A few years before his death, Joseph Fourier (1768–1830), a French mathematician, was the first scientist to realize that the atmosphere absorbs some of the radiation emanating from the ground; and in 1856, Eunice Foote, an American scientist and inventor, was the first author to link (briefly but clearly) CO_2 with global warming.[41] Five years later, John Tyndall (1820–1893), an English physicist, explained that water vapor is the most important absorber of the outgoing radiation, which means that "every variation of this constituent must produce a change in climate"—and he added that "similar remarks would apply to the carbonic acid diffused through the air."[42] Concise but clear, when rephrased in modern parlance it says: increases in CO_2 concentration must produce rising atmospheric temperature.

That was in 1861, and before the end of the century Svante Arrhenius (1859–1927), a Swedish chemist and an early Nobelian, published the first calculations of increased global surface temperature arising from the eventual doubling of preindustrial atmospheric CO_2.[43] His paper also noted that global warming will be felt least in the tropics and most felt in the polar regions, and that it will reduce temperature differences between night and day. Both of these conclusions have been confirmed. The Arctic warms up faster, but the simplest explanation (with the melting of snow and ice, the share of reflected radiation declines sharply, leading to higher warming) is only a part of a complex process that includes changes in clouds and water vapor and energy transport to the poles through large weather systems.[44] Nighttime temperatures are increasing faster than the daytime averages mainly because the boundary layer (the atmosphere just above the ground) is very thin—mere hundreds of meters—during the night, compared to several kilometers during the day, and hence it is more sensitive to warming.[45]

In 1908 Arrhenius provided a fairly accurate estimate of climate sensitivity, the measure of global warming arising from a doubling of atmospheric CO_2 level: "Any doubling of the percentage of carbon dioxide in the air would raise the temperature of the earth's surface by 4°C."[46] In 1957, three decades before the sudden surge of interest in global warming, Roger Revelle, an American oceanographer, and Hans Suess, a physical chemist, appraised the process of mass-scale fossil fuel combustion in its correct evolutionary terms: "Thus human beings are now carrying out a large scale geophysical experiment of a kind that could not have happened in the past nor be reproduced in the future. Within a few centuries we are returning to the atmosphere and oceans the concentrated organic carbon stored in sedimentary rocks over hundreds of millions of years."[47]

I cannot imagine what other phrasing could have better conveyed the unprecedented nature of this new reality. Just a year later, in response to this concern, the measurement of background CO_2 concentrations began at Mauna Loa in Hawai'i and at the South Pole, and they immediately showed constant and fairly predictable annual increases, from 315 ppm in 1958 to 346 ppm by 1985.[48] And in 1979, a report by the National Research Council put the theoretical value of climate sensitivity (including water vapor feedback) at 1.5°C–4.5°C, which means that the estimate offered by Arrhenius in 1908 was well within that range.[49]

The late-1980s "discovery" of carbon dioxide–induced global warming thus came more than a century after Foote and Tyndall made the link clear, nearly four generations after Arrhenius published a good quantitative estimate of the possible global warming effect, more than a generation after Revelle and Suess warned about an unprecedented and unrepeatable planet-wide geophysical experiment, and a decade after modern confirmation of climate sensitivity. Clearly, we did not have to wait for new computer models or for the establishment of an international bureaucracy to be aware of this change and to think about our responses.

How little fundamental difference these efforts have made is perhaps best illustrated by the latest estimates of a key global warming metric, climate sensitivity. The IPCC's fifth assessment report, published more than a century after Arrhenius offered the value of 4°C, concluded that it is extremely unlikely that the sensitivity is less than 1°C and very unlikely that it is above 6°C, with the likely range between 1.5°C and 4.5°C, the same as the 1979 National Research Council report.[50] And in 2019, a comprehensive assessment of Earth's climate sensitivity (using multiple lines of evidence) narrowed the most likely response to between 2.6°C and 3.9°C.[51] This means that climate sensitivity is extremely unlikely to be so low that it could prevent substantial warming (in excess of 2°C) by the time the atmospheric CO_2 concentration rises to about 560 ppm, twice the preindustrial level.

And yet, so far, the only effective, substantial moves toward decarbonization have not come from any determined, deliberate, targeted policies. Rather, they have been by-products of general technical advances (higher conversion efficiencies, more nuclear and hydro generation, less wasteful processing and manufacturing procedures) and of ongoing production and management shifts (switching from coal to natural gas; more common, less energy-intensive, material recycling) whose initiation and progress had nothing to do with any quest for reduced greenhouse gas emissions.[52] And, as already noted, the global impact of the recent turn toward decarbonizing electricity generation—by installing solar PV panels and wind turbines—has been completely negated by the rapid rise of greenhouse gas emissions in China and elsewhere in Asia.

Oxygen, water, and food in a warmer world

We know where we stand. Because of the rising concentrations of greenhouse gases, the planet has been, for generations, re-radiating slightly less energy that it has been receiving from the Sun. By 2020, the net value of this difference was about 2 watts per square meter compared to the 1850 baseline.[53] Because oceans have an enormous capacity to absorb atmospheric heat, it takes a long time to raise the average temperature of the lower atmosphere by an appreciable margin. During the late 2010s, after a couple of centuries of accelerated burning of fossil fuels, the temperature averaged across global land and ocean surfaces was almost 1°C above the 20th-century mean. It has been documented on all continents, but it has not been evenly distributed: as Arrhenius rightly predicted, higher latitudes have seen much higher average increases than the mid-latitudes or the tropics.

In terms of the global average, the five warmest years in the past 140 years have happened since 2015, and 9 of the 10 warmest years have been experienced since 2005.[54] There have been many consequences of this global change, ranging from the earlier flowering of Kyoto's cherry trees and earlier vintages of French wines to worrisome new temperature records during summer heat waves and the melting of high-mountain glaciers.[55] And (not surprisingly, given the ease of playing with many computer models) there is now an even more extensive literature predicting what is to come. So, circling back to the three existential fundamentals, what is the outlook for oxygen, water, and food supply on a warmer Earth?

Oxygen's atmospheric concentration is not affected by any slight greenhouse gas–driven changes in temperature, but it has been marginally declining because of the principal anthropogenic cause of global warming: the burning of fossil fuels. Their combustion has recently been removing about 27 billion tons of oxygen a year from the atmosphere.[56] Net annual decline of atmospheric oxygen (also taking into account its losses due to wildfires and livestock respiration) was put at about 21 billion tons at the beginning of the 21st century—that is, less than 0.002 percent of the existing concentration a year.[57] Direct measurements of atmospheric O_2 concentrations confirm these tiny losses: recently they have been about 4 ppm, and as there are nearly 210,000 oxygen molecules among every 1 million molecules of air, this translates to an annual decline of 0.002 percent.[58]

At this rate it would take 1,500 years (roughly equal to the time that has elapsed since the demise of the Western Roman Empire) to lower oxygen's atmospheric level by 3 percent—but in terms of actual O_2 concentrations, that is merely the equivalent of moving from New York City (at sea level) to Salt Lake City (1,288 meters above sea level). Another extreme—and completely theoretical—calculation shows that, even should we burn all of the world's known reserves of all fossil fuels (coal, crude oil, and natural gas: an impossibility due to the prohibitive costs of extracting these fuels from mostly marginal deposits), atmospheric O_2 concentration would be reduced by just 0.25 percent.[59]

Unfortunately, for hundreds of millions of people, breathing is made difficult for many reasons—ranging from pollen allergens to outdoor urban and indoor (from cooking) rural air pollution—but there is no risk of impaired breathing caused by any conceivable decline of atmospheric oxygen consumed by forest fires or by fossil fuel combustion. Moreover, no access to a vital natural resource is so equitable: whatever the local level of air pollutants, at the same altitude anywhere in the world an identical concentration of oxygen is freely available to anybody, and populations living at high altitudes, in places like Tibet and the Andes, have shown many remarkable adaptations (above all, elevated hemoglobin concentrations) to lower oxygen concentrations.[60]

All that is to say, we should not worry about oxygen. However, we must be concerned about the future of the water supply. Many regional, national, and global models have examined future water availability. They assume different degrees of global warming, and while the worst-case scenarios offer a generally deteriorating outlook, there are substantial uncertainties depending on necessary assumptions about population growth and therefore water demand. With a warming of up to 2°C, populations exposed to increased, climate change–induced water scarcity may be as low as 500 million and as high as 3.1 billion.[61] Per capita water supply will be decreasing worldwide, but some major river basins (including La Plata, the Mississippi, Danube, and Ganges) will remain well above the scarcity level, while some already water-scarce river basins will see further deterioration (perhaps most notably Turkey and Iraq's Tigris-Euphrates and China's Huang He).[62]

But most studies concur that demand-driven freshwater scarcity will have a much greater impact than the shortages induced by climate change. As a result, our best option for dealing with future water supply is to manage demand, and one of the best large-scale examples of this working is the recent history of the US's reducing per capita water usage.[63] In 2015, overall US water consumption was less than 4 percent higher than it was in 1965—but during the intervening 50 years the country's population increased by 68 percent, its GDP (in constant monies) more than quadrupled, and irrigated farmland increased by about 40 percent. This means that average per capita water use decreased by nearly 40 percent, that the water intensity of the US economy (units of water per unit of constant GDP) declined by 76 percent, and, as the total volume of water used for irrigation was actually slightly lower in 2015, that applications per unit of farmland declined by nearly a third. There are, of course, physical limits to the further reduction in all of these uses of water, but the US experience shows that the gains can go far beyond marginal ones.

The shortage of drinking water can be alleviated by desalination—the removal of dissolved salts from seawater by techniques ranging from solar distillation to the use of semipermeable membranes. This option has become more common in many water-short countries (there are some 18,000 desalination plants around the world) but the costs are well above those of freshwater supplied from reservoirs or from recycling.[64] The volumes of water needed for crops are of a much higher magnitude, and most of the world's food production will continue to depend on rainfall. Will there be enough of it in the coming warmer world?

Photosynthesis is always an extremely lopsided exchange of internal water (inside a leaf) for external CO_2 (in the atmosphere). Anytime a plant opens its stomata (located on the underside of leaves) to import sufficient carbon for its photosynthesis, it loses large amounts of water. For example, transpiration efficiency (biomass produced per unit of water used) of wheat (whole plant) is 5.6–7.5 grams per kilogram, and this translates to about 240–330 kilograms of water per kilogram of harvested grain.[65]

Global warming will, inevitably, intensify the water cycle because higher temperatures will increase evaporation. As a result, there will be, overall, more precipitation and hence more water available for capture, storage, and use.[66] But more precipitation in general will not mean more precipitation everywhere, nor—a no less important consideration—more precipitation when it is most needed. As with many other changes associated with a warmer climate, enhanced precipitation will be unevenly distributed. Some regions will be receiving less than today; others (including the Yangtze basin, home to most of China's large population) significantly more, and this increase is expected to bring a slight reduction in the number of people residing in highly water-stressed environments.[67] But many places with more precipitation will get it in a more irregular manner, in the form of less frequent but heavier—even catastrophic—rain or snow events.

A warmer atmosphere will also enhance water loss from plants (evapotranspiration), but that does not mean that crops and forests will wilt as they lose water. A rising atmospheric level of CO_2 means that the water required per unit of yield will decrease in a warmer and CO_2-richer biosphere. This effect has already been measured in some crops, and wheat and rice (staple grains which rely on the most common photosynthetic pathway) will increase their water use efficiency more than corn or sugar cane (which use a less common, but inherently more efficient, pathway).[68] This means that, in some regions, wheat and other crops could yield as much or more than today, even if the precipitation they receive is reduced by 10–20 percent.

At the same time, global food production is also a significant source of trace gases that contribute to global warming, mostly CO_2 from conversions of forests and pastures to fields (still ongoing above all in South America and Africa), and methane emissions from ruminant livestock.[69] But this reality also presents opportunities for improvements and adjustments. Crops could be grown in ways that increase organic matter in soils and hence their carbon storage (by reducing or eliminating annual plowing), and methane emissions from livestock could be lowered by eating less beef. My calculations show that in the future—by lowering the share of beef and raising the share of pork, chicken meat, eggs, and dairy products, by more efficient feeding, and by better use of crop residues and food processing by-products—we could match recent global meat output while greatly limiting livestock's environmental impact, including its share of methane emissions.[70]

More broadly, a recent study asked if it is possible to feed the future population of 10 billion people (expected shortly after 2050) within four terrestrial planetary boundaries—in other words, doing so without taking the Earth and its inhabitants to the brink in terms of transgressing the limits on the biosphere's integrity, land and freshwater use, and nitrogen flow. Not surprisingly, the study concluded that if all of those boundaries were strictly respected, the global food system could supply balanced daily diets (about 2,400 kilocalories per capita) to no more than 3.4 billion people—but that 10.2 billion could be supported with the redistribution of cropland, better water and nutrient management, food waste reduction, and dietary adjustments.[71]

Informed looks at the three existential necessities of life—breathing, drinking, and eating—concur: there should be no unavoidable apocalypse by 2030 or 2050. Oxygen will remain abundant. Concerns about water supply will increase in many regions, but we have the knowledge and we should be able to mobilize the means needed to avert any mass-scale life-threatening shortages. And we should not only maintain but improve average per capita food supply in low-income countries, while reducing excessive production in affluent nations. However, these actions would only reduce, not eliminate, our reliance on direct and indirect fossil fuel subsidies in the production of food for the global population (see chapter 2). And, as I explained in the first chapter, moving away from fossil fuels cannot be done rapidly. This means that, for decades to come, their combustion will remain the principal driver of global climate change. How will this affect the long-term trend of global warming?

Uncertainties, promises, and realities

The combination of scientific advances and improving technical capabilities means that we now approach any complex process that involves the intricate interplay of natural factors and human actions with the advantages of considerable, and steadily expanding, understanding. At the same time, we also have to reckon with uncomfortable degrees of ignorance, and with persistent uncertainties that make any resolute response so much more difficult. If a reminder of this fundamental reality was needed, then the spread and the consequences of COVID-19 provided many sobering global lessons.

We were unprepared—to a degree even those of us who expected major problems found surprising—for an event whose near-imminent occurrence could have been forecast with 100 percent certainty: in 2008 I did so unequivocally in my book on global catastrophes and trends, getting even the timing right.[72] Although we almost immediately identified the full genetic makeup of this new pathogen, national public policy responses to its spread ranged from largely business as usual (Sweden) to draconian (but belated) countrywide shutdowns (Italy, Spain), and from early dismissals (the US in February 2020) to early successes that turned into later problems (Singapore).[73]

And yet, fundamentally, this is a self-limiting natural phenomenon that we have experienced on a global scale three times since the late 1950s: even without any vaccines, every viral pandemic eventually subsides once the pathogen infects relatively large numbers of people or once it mutates into a less virulent form. In contrast, global climate change is an extraordinarily complex development whose eventual outcome depends on far-from-perfectly understood interactions of many natural and anthropogenic processes. As a result, we will need, for decades to come, more observations, more studies, and far better climate models in order to get more accurate appraisals of long-term trends and of the most likely outcomes.

To believe that our understanding of these dynamic, multifactorial realities has reached the state of perfection is to mistake the science of global warming for the religion of climate change. At the same time, we do not need an endless stream of new models in order to take effective actions. There are enormous opportunities for reducing energy use in buildings, transportation, industry, and agriculture, and we should have initiated some of these energy-saving and emissions-reducing measures decades ago, regardless of any concerns about global warming. Quests to avoid unnecessary energy use, to reduce air pollution and water, and to provide more comfortable living conditions should be perennial imperatives, not sudden desperate actions aimed at preventing a catastrophe.

Most remarkably, we have largely ignored taking steps that could have limited the long-term impacts of climate change and that should have been taken even in the absence of any global warming concerns because they bring long-term savings and provide more comfort. And as if that were not enough, we have deliberately introduced and promoted the diffusion of new energy conversions that have boosted the consumption of fossil energies and hence further intensified CO_2 emissions. The best examples of these omissions and commissions are the indefensibly inadequate building codes in cold-climate countries and the worldwide adoption of SUVs.

Because our houses are around for a long time (a well-built, properly maintained wood-frame North American house with a concrete foundation can last for more than 100 years), with proper wall insulation, triple window panes, and highly efficient heating furnaces they represent a unique opportunity for long-lasting energy (and hence carbon emissions) savings.[74] In 1973, when OPEC quintupled the world price of crude oil, most buildings in Europe, North America, and North China had just single-pane windows; in Canada triple panes will not be mandated before 2030, and Manitoba was the first province to require high-efficiency (>90 percent) natural gas–fired furnaces in 2009, decades after such options became commercially available.[75] Would it not be interesting to know how many delegates to global warming meetings coming from cold climates have triple panes filled with inert gas, super-insulated walls, and 97 percent–efficient gas furnaces? Analogically, how many people in hot climates have properly sealed rooms so that their poorly installed and inefficient window air conditioners don't waste cool air?

SUV ownership began to rise in the US during the late 1980s, it eventually diffused globally, and by 2020 the average SUV emitted annually about 25 percent more CO_2 than a standard car.[76] Multiply that by the 250 million SUVs on the road in 2020, and you will see how the worldwide embrace of these machines has wiped out, several times over, any decarbonization gains resulting from the slowly spreading ownership (just 10 million in 2020) of electric vehicles. During the 2010s, SUVs became the second-highest cause of rising CO_2 emissions, behind electricity generation and ahead of heavy industry, trucking, and aviation. If their mass public embrace continues, they have the potential to offset any carbon savings from the more than 100 million electric vehicles that might be on the road by 2040!

The second chapter of this book detailed the high energy cost of modern food production and noted indefensibly high levels of food waste: clearly, this combination presents many opportunities for reducing not only CO_2 emissions but also CH_4 emissions from rice cultivation and ruminant livestock and N_2O emissions from the excessive application of nitrogen fertilizer—as well as the emissions from questionable food trade. Is it necessary to airlift blueberries from Peru to Canada in January, and green beans from Kenya to London? The vitamin C and roughage these foodstuffs provide can be secured from many other sources with much lower carbon footprints. And could not we, with our immense data-processing capabilities, price food better and more flexibly in order to make a major dent in the 30–40 percent waste rate? Why not do what can be done, profitably and immediately, rather than waiting for more modeling exercises?

The list of what we have not—but could have—done is long. And what have we done to avert, or to reverse, the unfolding environmental change in the three decades since global warming became a dominant topic of modern discourse? The data are clear: between 1989 and 2019 we increased global anthropogenic greenhouse gas emissions by about 65 percent. Even when we deconstruct this global mean, we see that affluent countries like the US, Canada, Japan, Australia, and those in the EU, whose per capita energy use was very high three decades ago, did reduce their emissions, but only by about 4 percent, while Indian emissions quadrupled and Chinese emissions rose 4.5 times.[77]

The combination of our inaction and of the extraordinarily difficult nature of the global warming challenge is best illustrated by the fact that three decades of large-scale international climate conferences have had no effect on the course of global CO_2 emissions. The UN's first conference on climate change took place in 1992; annual climate change conferences began in 1995 (in Berlin) and included much publicized gatherings in Kyoto (1997, with its completely ineffective agreement), Marrakech (2001), Bali (2007), Cancún (2010), Lima (2014), and Paris (2015).[78] Clearly, the delegates love to travel to scenic destinations with hardly any thought of the dreaded carbon footprint generated by this global jetting.[79]

In 2015, when about 50,000 people flew to Paris in order to attend yet another conference of the parties at which they were to strike, we were assured, a "landmark"—and also "ambitious" and "unprecedented"—agreement, and yet the Paris accord did not (could not) codify any specific reduction targets by the world's largest emitters, and it would, even if all voluntary non-binding pledges were honored (something utterly improbable), result in a 50 percent *increase* of emissions by 2050.[80] Some landmark.

These meetings could never have stopped either the expansion of China's coal extraction (it more than tripled between 1995 and 2019, to nearly as much as the rest of the world combined) or the just-noted worldwide preference for massive SUVs, and they could not have dissuaded millions of families from purchasing—as soon as their rising incomes allowed—new air conditioners that will work through the hot humid nights of monsoonal Asia and hence will not be energized by solar electricity anytime soon.[81] The combined effect of these demands: between 1992 and 2019, the global emissions of CO_2 rose by about 65 percent; those of CH_4 by about 25 percent.[82]

What can we do during the coming decades? We must start with the recognition of fundamental realities. We used to consider a 2°C increase in average global temperature as a relatively tolerable maximum, but in 2018 the IPCC lowered that to just 1.5°C—but by 2020 we have already added two-thirds of that maximum preferable temperature rise. Moreover, in 2017 an assessment that considered the capacity of the oceans to absorb carbon, the planet's energy imbalances, and the behavior of fine particles in the atmosphere concluded that the committed global warming (arising from past emissions and becoming a reality even if all new emissions ceased instantly) had already added up to 1.3°C and hence it would require only an additional 15 years of new emissions to surpass 1.5°C.[83] The latest analysis of these combined effects concluded that we are already committed to global warming of 2.3°C.[84]

As always, these conclusions have their own margins of error, but it seems highly likely that the proverbial horse of 1.5°C warming has already bolted. Even so, many institutions, organizations, and governments are still theorizing about keeping it inside the broken corral. The IPCC report on 1.5°C warming offers a scenario based on such a sudden and persistent reversal of our reliance on fossil fuels that the global emissions of CO_2 would be halved by 2030 and eliminated by 2050[85]—and other scenario-builders are now offering detailed suggestions on how to achieve a rapid end to the fossil carbon era. Computers make it easy to construct many scenarios of rapid carbon elimination—but those who chart their preferred paths to a zero-carbon future owe us realistic explanations, not just sets of more or less arbitrary and highly improbable assumptions detached from technical and economic realities and ignoring the embedded nature, massive scale, and enormous complexity of our energy and material systems. Three recent exercises provide excellent illustrations of these flights of fancy unencumbered by real-world considerations.

Wishful thinking

The first scenario, prepared mostly by EU researchers, assumes that average global per capita energy demand in 2050 will be 52 percent lower than it was in 2020. Such a drop would make it easy to keep the global temperature rise below 1.5°C (that is, if we still believe that such a thing is possible).[86] Of course—and as I will reiterate in the final chapter—when constructing long-range scenarios, we can plug in any arbitrary assumptions in order to meet preconceived outcomes. But how do the assumptions of this scenario align with the recent past?

Cutting per capita energy demand by half in three decades would be an astonishing accomplishment given the fact that over the previous

30 years global per capita energy demand rose by 20 percent. The projection assumes that the much lower demand for energy will arise from a combination of moving away from owning goods, the digitalization of daily life, and a rapid diffusion of technical innovations in converting and storing energies.

The suggested first driver of disappearing demand (owning less) is an academic belief for which there is very little evidence, as all major categories of personal consumption—measured by annual household expenditures—have been rising even in affluent countries. In already highly saturated markets with already congested traffic, the EU's car ownership per 1,000 people rose by 13 percent between 2005 and 2017, and during the past 25 years it was up about 25 percent in Germany and 20 percent in France.[87] Reduced demand and gradual declines in ownership are desirable and likely; halving the demand is an arbitrary and unlikely target.

More importantly, the proponents of this unrealistic scenario allow merely a factor-of-two increase across all modes of mobility during the next three decades in what they call the Global South (a common but highly inaccurate designation of low-income nations, mostly in Asia and Africa), and a factor-of-three increase in the ownership of consumer goods. But in the China of the past generation, growth has been on an entirely different scale: in 1999 the country had just 0.34 cars per 100 urban households, in 2019 the number surpassed 40. That is a more than 100-fold relative increase in only two decades.[88] In 1990, 1 out of every 300 urban households had an air-conditioning window unit; by 2018 there were 142.2 units per 100 households: a more than 400-fold rise in less than three decades. Consequently, even if those countries whose standard of living is today where China's was in 1999 were to achieve only a tenth of China's recent growth, they would experience a 10-fold increase of car ownership and a 40-fold increase in air conditioners. Why do the prescribers of the low-energy-demand scenario think that today's Indians and Nigerians do not want to narrow the gap that separates them from China's material ownership?

Not surprisingly, the latest global production gap report—an annual publication highlighting the discrepancy between fossil fuel production planned by individual countries and the global emission levels necessary to limit warming to 1.5°C or 2°C—does not show any commitments to plunging trend lines; just the opposite, in fact.[89] In 2019, the major consumers of fossil energies were aiming to produce 120 percent more fuels by 2030 than would be consistent with limiting global warming to 1.5°C, and whatever the eventual effect of the COVID-19 pandemic, the resulting decline of consumption will be both temporary and too small to reverse the general trend.

In the second scenario conforming to the goal of complete decarbonization by 2050, a large group of energy researchers at Princeton University have charted the required shifts in the US.[90] The Princeton scenario-builders recognize that it will be impossible to eliminate all fossil fuel consumption and that the only way to achieve net-zero emissions is to resort to what they label the "fourth pillar" of their overall strategy—to mass-scale carbon capture and storage of emitted CO_2—and their calculation requires the removal of 1–1.7 gigatons of the gas per year. When compared on a volume equivalent basis, that would necessitate the creation of an entirely new gas-capture-transportation-storage industry that every year would have to handle 1.3–2.4 times the volume of current US crude oil production, an industry that took more than 160 years and trillions of dollars to build.

The majority of this carbon storage is to take place along the Texas Gulf Coast, and this would require building about 110,000 kilometers of new CO_2 pipelines, necessitating an utterly unprecedented speed in the planning, permitting, and constructing of such extensive links in a society famous for its litigiousness and NIMBY resistance.[91] At the same time, additional monies would have to be spent on dismantling the existing transmission infrastructure of the US oil and gas industry. Given the rich historical experience with massive long-term cost overruns, any cost estimates for expenditure over the next three decades cannot be trusted even as far as their order of magnitude is concerned.

Achieving complete decarbonization by 2050 is a tame goal compared to the third scenario, which extends the goals of the US Green New Deal (introduced in the US Congress in 2019) to 143 countries and outlines how at least 80 percent of global energy supply will be decarbonized by 2030 thanks to renewable wind, water, and solar (WWS) energy, the supply of which will reduce overall needs by 57 percent, financial costs by 61 percent, and social (health and climate) costs by 91 percent: "Thus, 100 percent WWS needs less energy, costs less, and creates more jobs than current energy."[92] There is no shortage of media, celebrities, and bestselling authors repeating, supporting, and amplifying these claims, ranging (no surprise) from *Rolling Stone* to the *New Yorker*, and from Noam Chomsky (who adds energy as his latest field of expertise) to Jeremy Rifkin, who believes that without such an intervention our fossil-fueled civilization will collapse by 2028.[93]

If true, these claims and their enthusiastic endorsements raise the obvious question: why should we worry about global warming? Why be frightened by the idea of early planetary demise, why feel compelled to join Extinction Rebellion? Who could be against solutions that are both cheap and nearly instantly effective, that will create countless well-paying jobs and ensure care-free futures for coming generations? Let us all just sing from these green hymnals, let us follow all-renewable prescriptions and a new global nirvana will arrive in just a decade—or, if things get a bit delayed, by 2035.[94]

Alas, a close reading reveals that these magic prescriptions give no explanation for how the four material pillars of modern civilization (cement, steel, plastic, and ammonia) will be produced solely with renewable electricity, nor do they convincingly explain how flying, shipping, and trucking (to which we owe our modern economic globalization) could become 80 percent carbon-free by 2030; they merely assert that it could be so. Attentive readers will remember (see chapter 1) that during the first two decades of the 21st century Germany's unprecedented quest for decarbonization (based on wind and solar) succeeded in boosting the shares of wind- and solar-generated electricity to more than 40 percent, but it lowered the share of fossil fuels in the country's primary energy use only from about 84 percent to 78 percent.

What miraculous options will be available to African nations now relying on fossil fuels to supply 90 percent of their primary energy, in order to drive their dependence to 20 percent within a decade while also saving enormous sums of monies? And how will China and India (both countries are still expanding their coal extraction and coal-fired generation) suddenly become coal-free? But these specific critiques of published rapid-speed transformation narratives are really beside the point: it makes no sense to argue with the details of what are essentially the academic equivalents of science fiction. They start with arbitrarily set goals (zero by 2030 or by 2050) and work backwards to plug in assumed actions to fit those achievements, with actual socioeconomic needs and technical imperatives being of little, or no, concern.

Reality thus presses in from both ends. The sheer scale, cost, and technical inertia of carbon-dependent activities make it impossible to eliminate all of these uses in just a few decades. As I detailed in the chapter on energy, we cannot sever that dependence so rapidly, and every realistic long-term forecast concurs: most notably, even the IEA's most aggressive decarbonization scenario has fossil fuels supplying 56 percent of the global primary energy demand by 2040. Similarly, the enormous scale and cost of material and energy demands make it impossible to resort to direct air capture as a decisive component of rapid global decarbonization.

But we can make a great deal of difference, not by pretending to follow unrealistic and arbitrary goals: all too obviously, history does not unfold as a computerized academic exercise with major achievements falling on years ending with zero or five; it is full of discontinuities, reversals, and unpredictable departures. We can proceed fairly fast with the displacement of coal-fired electricity by natural gas (when produced and transported without significant methane leakage, it has a substantially lower carbon intensity than coal) and by expanding solar and wind electricity generation. We can move away from SUVs and accelerate mass-scale deployment of electric cars, and we still have large inefficiencies in construction, household, and commercial energy use that can be profitably reduced or eliminated. But we cannot instantly change the course of a complex system consisting of more than 10 billion tons of fossil carbon and converting energies at a rate of more than 17 terawatts, just because somebody decides that the global consumption curve will suddenly reverse its centuries-long ascent and go immediately into a sustained and relatively fast decline.

Models, doubts, and realities

Why is it that some scientists keep on charting such arbitrarily bending and plunging curves leading to near-instant decarbonization? And why are others promising the early arrival of technical superfixes that will support high standards of living for all humanity? And why are these wishful offerings taken so often for reliable previsions and are readily believed by people who would never try to question their assumptions? I will have more to say about this in the closing chapter, but here are some observations related to the now-so-dominant concern of global environmental change.

De omnibus dubitandum (Doubt everything) must be more than a durable Cartesian quote; it must remain the very foundation of the scientific method. Recall how I opened this chapter with a list of nine planetary boundaries whose transgressions imperil our biospheric wellbeing? Keeping them within safe confines seems to be an obvious conclusion because they identify the most important, perennial, existential concerns—and yet a list prepared 40 years ago would have been very different. Acid rain (or, more correctly, acidifying precipitation) would have been, most likely, its top item, because a broad consensus of the early 1980s saw it as the leading environmental problem.[95]

Stratospheric ozone depletion would have been absent because the infamous Antarctic ozone hole was discovered only in 1985; and, if listed at all, anthropogenic climate change and associated ocean acidification would have been near the bottom.[96] And even when focusing on such perennial concerns as land use changes (dominated by deforestation), loss of biodiversity (from iconic pandas and koalas to bee colonies and sharks), and freshwater supply, our concerns have evolved considerably, becoming more acute in some ways (we now worry more about withdrawals of underground water and about excessive nutrients creating coastal dead zones) and less pressing in others (perhaps most notably, forests have made substantial comebacks, not only in all affluent countries but also in China).[97]

When looking ahead, we must regain a critical perspective when dealing with all models exploring environmental, technical, and social complexities. There are no limits to assembling such models or, as fashionable lingo has it, constructing narratives. Their authors can choose, as so many recent climate models have done recently, excessive assumptions about future energy use, and they can end up with very high rates of warming that generate news headlines about hellish futures.[98] Taking the opposite approach, other modelers can posit 100 percent inexpensive thermonuclear electricity or cold fusion by 2050, or, alternatively, they can allow for the unlimited expansion of fossil fuel combustion, because their model deploys miraculous techniques that will not only remove any volume of CO_2 from the atmosphere but recycle it as a feedstock for synthesizing liquid fuel—all at a steadily declining cost.

Of course, they just march along with the new tech crowd whose naivete compares every technical change to recent developments in electronics, and above all to mobile phones. Here is how a green energy CEO put it in 2020: "Do you remember how we transformed telephony from fixed-line phones to mobile phones, television from watching whatever was on TV to whatever we fancied, from buying newspapers to customising our news feeds? The people-led, tech-powered energy revolution is going to be just the same."[99] How could changing a device (landline to mobile) whose reliable use depends on a massive, complex, and highly reliable system of electricity generation (dominated by thousands of large fossil-fueled, hydro, and nuclear power plants), transformation, and transmission (encompassing hundreds of thousands of kilometers of national and even continental-scale grids) be the same as changing the entire underlying system?

Much of this unmoored thinking comes out as intended—ranging from scary to wonderful—and I can see why many people are taken in either by these threats or by unrealistic suggestions. Only the imagination limits these assumptions: they range from fairly plausible to patently delusionary. This is a new scientific genre where heavy doses of wishful thinking are commingled with a few solid facts. All of these models should be seen mainly as heuristic exercises, as bases for thinking about options and approaches, never to be mistaken for prescient descriptions of our future. I wish this admonition would be as obvious, as trivial, and as superfluous as it seems!

Regardless of the perceived (or modeled) severity of global environmental challenges, there are no swift, universal, and widely affordable solutions to tropical deforestation or biodiversity loss, to soil erosion or to global warming. But global warming presents an uncommonly difficult challenge precisely because it is a truly global phenomenon, and because its largest anthropogenic cause is the combustion of fuels that constitute the massive energetic foundations of modern civilization. As a result, non-carbon energies could completely displace fossil carbon in a matter of one to three decades ONLY if we were willing to take substantial cuts to the standard of living in all affluent countries and deny the modernizing nations of Asia and Africa improvements in their collective lots by even a fraction of what China has done since 1980.

Still, major reductions in carbon emissions—resulting from the combination of continued efficiency gains, better system designs, and moderated consumption—are possible, and a determined pursuit of these goals would limit the eventual rate of global warming. But we cannot know to what extent we will succeed by 2050, and thinking about 2100 is truly beyond our ken. We can outline extreme cases, but in just a few decades the fan of possible outcomes becomes too wide and, in any case, the progress of any eventual decarbonization is contingent not only on our deliberate remedial actions but also on unpredictable intervening changes in national fortunes.

Was there a single climate modeler who predicted in 1980 the most important anthropogenic factor driving global warming over the past 30 years: the economic rise of China? At that time even the best models, all being direct descendants of global atmospheric circulation models developed during the 1960s, had no way of reflecting unpredictable shifts in national fortunes, and they also ignored the interactions between the atmosphere and the biosphere. That did not make these models useless: they assumed the continued global growth of greenhouse gas emissions and, in general, they were fairly accurate in predicting the rate of global warming.[100]

But a good estimate of the overall rate is only the beginning. To use, once again, the COVID-19 analogy, this is akin to making a forecast in 2010 that—based on the last three pandemics and adjusted for a larger population—global deaths during the first year of the next global pandemic would be about 2 million.[101] That would be very close to the actual total—but would that forecast (correctly assuming, based on many precedents, that the pandemic would start in China) also assign only 0.24 percent of those deaths (in absolute terms, less than in Greece or Austria) to China, a country with nearly 20 percent of the global population—and nearly 20 percent to the US, a far richer and (it certainly believes this about itself) far more competent country with less than 5 percent of the global population?

And, even more incredibly, would it predict that the highest mortalities would be concentrated in the most affluent Western economies, those that boast about their state-delivered advanced health care? In March 2021, as the pandemic entered officially its second year (the WHO proclaimed it on March 11, 2020, although the infection had been spreading in China at least since December 2019), all the top 10 countries with the highest cumulative mortalities (above 1,500 per million; or 1.5 out of every 1,000 people died of COVID-19) were in Europe, including six EU members and the UK. And who could predict that the US rate (also above 1,500) would be two orders of magnitude above China's 3 deaths per 1 million?[102] Obviously, even a highly accurate forecast of total interim COVID-19 mortality would provide no specific guidance for formulating the best national responses.

Analogically, the post-1980 rise of China (as well as India) has changed the circumstances of any response to rising global trace gas emissions. In 1980, four years after Mao Zedong's death, China's per capita economic product was less than one-quarter of the Nigerian mean; there were no private passenger cars; only the top Communist Party leaders living in the seclusion of Zhongnanhai (the former imperial garden within the Forbidden City, now the central headquarters of the Communist Party) had air conditioning; and China produced just 10 percent of global CO_2 emissions.[103]

By 2019 China was, in terms of purchasing power, the world's largest economy; its per capita GDP was five times the Nigerian mean; the country was the world's largest producer of cars; half of all urban households had two window-mounted air-conditioning units; the length of its rapid train network surpassed the combined length of all the EU's links; and about 150 million of its citizens had traveled overseas. The country also emitted 30 percent of the world's CO_2 from fossil fuels. In contrast, the combined emissions of the USA and EU28 fell from 60 percent of the world total in 1980 to 23 percent by 2019 and (because of slow rates of economic growth; aging, even declining, populations; and large-scale offshoring of industrial production to Asia) their combined share is highly unlikely to rise ever again.

Looking ahead, most of the power to enact meaningful change will lie more and more in the modernizing economies of Asia: excluding high-income and low- or no-population growth Japan, South Korea, and Taiwan, the continent is now producing half of all emissions. And while the unfolding transformation of sub-Saharan Africa has been much slower, its combined population of about 1.1 billion will nearly double during the next 30 years, it will contain almost 50 percent more people than China (the country that all low-income economies wish to emulate), and a critical assessment of the continent's electricity future points to a high-carbon lock-in, with fossil-fueled generation dominant and with the share of non-hydro renewables remaining below 10 percent in 2030.[104]

The rise and fall of nations is not the only uncertainty concerning the progress and effects of global warming. Recent good news is that the world's forests have been a large and persistent carbon sink (storing more than they emit), locking away about 2.4 billion tons of carbon every year between 1990 and 2007, and satellite data for 2000–2017 indicate that one-third of the world's vegetated area has been greening (showing a significant increase in the average annual green leaf area, confirming that more carbon is now absorbed and stored) and only 5 percent has been browning (showing significant loss of leaves).[105] This effect has been particularly notable in intensively cultivated croplands in China and India, and in China the effect has also been seen in the country's expanding forests.

But the not-so-good news (you knew it was coming . . .) is that between 1900 and 2015 the biosphere lost 14 percent of its trees due to cutting and, no less significantly, tree mortality doubled during that time, with older (and taller) trees accounting for a larger share of this loss. The world's forests are getting younger and shorter, and hence are not able to store as much carbon as in the past.[106] Increased growth rates appear to be shortening trees' lifespans across almost all species and climates, and so the existence of substantial carbon sinks may be only transient.[107] And how many times have you heard that, inevitably, the first places to succumb to rising sea levels caused by global warming will be the low-lying coasts in general and island nations in the Pacific in particular?[108] And yet a recent analysis of four decades of shoreline change in all 101 islands in the Pacific atoll nation of Tuvalu (north of Fiji, east of the Solomon Islands) shows that the nation's land area has actually increased by nearly 3 percent.[109] Preconceived and rashly generalizing conclusions should always be avoided.

The evolution of societies is affected by the unpredictability of human behavior, by sudden shifts of long-lasting historical trajectories, by the rise and fall of nations, and is accompanied by our ability to enact meaningful changes. These realities affect many inherently complex (and far from satisfactorily understood) biospheric processes. And because they elicit often contradictory natural responses, such as forests being both sinks and sources of carbon, it is impossible to say with confidence where we will be—in terms of fossil fuel consumption, the pace of decarbonization, or environmental consequences—in 2030 or 2050.

Most notably, what remains in doubt is our collective—in this case global—resolve to deal effectively with at least some critical challenges. Solutions, adjustments, and adaptations are available. Affluent countries could reduce their average per capita energy use by large margins and still retain a comfortable quality of life. Widespread diffusion of simple technical fixes ranging from mandated triple windows to designs of more durable vehicles would have significant cumulative effects. The halving of food waste and changing the composition of global meat consumption would reduce carbon emissions without degrading the quality of food supply.

Remarkably, these measures are absent, or rank low, in typical recitals of coming low-carbon "revolutions" that rely on as-yet-unavailable mass-scale electricity storage or on the promise of unrealistically massive carbon capture and its permanent storage underground. There is nothing new about these exaggerated expectations.

In 1991 a well-known environmental activist wrote about "abating global warming for fun and profit."[110] If this promise would have even remotely resembled reality, we would not be faced, three decades later, with the increasing anguish of today's warming catastrophists. Similarly, we are now promised even more astonishing "disruptive" innovations and AI-driven "solutions." The reality is that any sufficiently effective steps will be decidedly non-magical, gradual, and costly. We have been transforming the environment on increasing scales and with rising intensity for millennia, and we have derived many benefits from these changes—but, inevitably, the biosphere has suffered. There are ways to reduce those impacts but the resolve to deploy them at required scales has been lacking, and if we start acting in a sufficiently effective manner (and this now requires doing so on a global scale) we will have to pay a considerable economic and social price. Will we, eventually, do so deliberately, with foresight; will we act only when forced by deteriorating conditions; or will we fail to act in a meaningful way?

7. Understanding the Future

Between Apocalypse and Singularity

"Apocalypse" comes (via Latin) from the ancient Greek ἀποκάλυψις. Literally, it means "uncovering." In the Christian context, the meaning shifted toward a prophesied unveiling, or revelation, of the second coming, and in modern usage the term has become synonymous with the end of life on Earth, the doomsday, or—to use another Greek biblical term—Armageddon.[1] Clear and unambiguously definitive.

Apocalyptic visions of the future—with assorted hells offered by major religions—have been strongly revived by modern promoters of doom, who have been pointing to rapid population growth, environmental pollution, or now, increasingly, to global warming as the sins that will transport us to the netherworld. In contrast, incorrigible techno-optimists continue the tradition of believing in miracles and the delivery of eternal salvation. It is not uncommon to read how artificial intelligence and deep learning systems will carry us all the way to the "Singularity." This comes from the Latin *singularis*, meaning "individual, unique, unmatched"—but in this chapter it refers to futurist Ray Kurzweil's notion of singularity, i.e. to the mathematical meaning of the term as a point in time at which a function assumes an infinite value.[2] He predicts that, come 2045, machine intelligence will have surpassed human intelligence, what he calls biological and nonbiological intelligence will merge, and machine intelligence will fill the universe at infinite speed.[3] This is the ultimate ascension. It will make colonization of the rest of the universe an inevitably effortless endeavor.

Long-range modeling of complex systems often relies on producing a fan of possible outcomes constrained by plausible extremes. Apocalypse and singularity offer two absolutes: our future will have to lie somewhere within that all-encompassing range. What has been so remarkable about modern anticipations of the future is how they have gravitated—despite all of the available evidence—toward one of these two extremes. In the past, this tendency toward dichotomy was often described as the clash of catastrophists and cornucopians, but these labels appear to be too timid to reflect the recent extreme polarization of sentiments.[4] And this polarization has been accompanied by a greater propensity for dated quantitative forecasts.

You see them everywhere, from cars (worldwide sales of electric passenger vehicles will reach 56 million by 2040) and carbon (the EU will have net-zero carbon emissions by 2050) to global flying (there will be 8.2 billion travelers by 2037).[5] Or so we're told. In reality, most of these forecasts are no better than simple guesses: any number for 2050 obtained by a computer model primed with dubious assumptions—or, even worse, by a politically expedient decision—has a very brief shelf life. My advice: if you would like a better understanding of what the future may look like, avoid these new-age dated prophecies entirely, or use them primarily as evidence of prevailing expectations and biases.

For generations, businesses and governments were the most common practitioners and consumers of forecasting, then academics joined the game in large numbers from the 1950s, and now anybody can be a forecaster—even without any mathematical skills—simply by using plug-in software or (as has been in vogue lately) by making baseless qualitative predictions. As in so many other cases of newly expanded endeavors (information flows, mass education), the quantity of modern forecasting has become inversely proportional to its quality. Many forecasts are nothing but the simplest extensions of past trajectories; others are the outcome of complicated interactive models which incorporate large numbers of variables and run using different assumptions each time (essentially the numerical equivalent of narrative scenarios); and some have hardly any quantitative component and are just wishful and exceedingly politically correct narratives.

Quantitative forecasts fall into three broad categories. The smallest includes forecasts that deal with processes whose workings are well known and whose dynamics are inherently restricted to a relatively confined set of outcomes. The second, and a much larger category, includes forecasts pointing in the right direction but with substantial uncertainties regarding the specific outcome. And the third category (I already described some of its recent energy and environmental specimens in the previous chapter) is that of quantitative fables: such forecasting exercises may teem with numbers, but the numbers are outcomes of layered (and often questionable) assumptions, and the processes traced by such computerized fairy tales will have very different real-world endings. Of course, their creators may defend the heuristic value of such exercises, while uninitiated users may exploit some of the conclusions to reinforce their own prejudices or to dismiss plausible alternatives.

Only the forecasts (projections, computer models) in the first category provide solid insights and good guidance, especially when looking only a decade or so ahead. Demographic projections in general, and fertility forecasts in particular, are among the best examples in this limited category. Take a country whose total fertility rate—that is, the number of children the average woman has in her lifetime—has been below the replacement level (an average of at least 2.1 children per woman is needed to replace the parents) for a generation and, moreover, retreated from 1.8 to 1.5 over the past decade. Such very low fertilities are unlikely to be reversed (no country has done so during the past three decades) to bring any substantial population increase within the next 10 years.[6] The most likely prospects are fertility that may recover a bit (from 1.5 to 1.7) or drop further (to 1.3). While it is impossible to pinpoint the value even in just 10 years, a forecast can offer a relatively narrow range of highly plausible outcomes. For example, the UN's 2019 population forecast for the year 2030 has Poland's total (37.9 million in 2020) declining to 36.9 million, with the low and high variant departing just ±2 percent from the mean, and (barring mass immigration that is unlikely in such an immigration-averse country) there is a very high probability that the actual 2030 count will be within that narrow range.[7]

In contrast, even short-term projections involving complex systems—those that reflect interplays of many technical, economic, and environmental factors, and which can be strongly affected by a number of arbitrary decisions such as unexpectedly generous government subsidies or new laws or sudden policy reversals—remain highly uncertain, and even near-term outlooks result in a broad range of possible outcomes. Forecasts for the worldwide adoption of electric passenger cars are an excellent recent example in this category.[8] Technical difficulties accompanying the introduction of personal electromobility have not been insurmountable, but the sector has been maturing much slower than its uncritical proponents began to claim years ago, while combustion engines keep improving their efficiency and will offer for years to come the advantages of lower initial cost, generations-long familiarity, and ubiquitous servicing.[9]

And while some countries have been aggressively promoting electric car ownership by offering generous subsidies or by mandating specific shares of new vehicles in the future, others have offered only minor or no help. As a result, past near-term forecasts for the worldwide electrification of road transport have almost uniformly overestimated the actual share: between 2014 and 2016 it was put as high as 8–11 percent by 2020, while the actual share was just 2.5 percent.[10] And by 2019, forecasts for electrics as a share of all vehicles on the road by the year 2030 differed by an order of magnitude, while the actual sales of internal combustion vehicles may outnumber those of the electrics for more than a decade.[11]

The third category of quantitative forecasts is the one that merits a closer look, because in retrospect many of them have not only failed to capture at least a proper order of magnitude but their claims and conclusions have turned out to be completely at odds with what actually happened. Remarkably, this is not only true of well-known historical prophecies ranging from the Bible to Nostradamus.[12] Many modern prophets have not done much better, yet with the rise of ubiquitous computing their ranks have swelled, and with the insatiable demand from the media for new bad news, their predictions and scenarios receive unprecedented distribution and (increasingly global) attention.

Failed predictions

Given the abundance of failed forecasts, their systematic recounting, be it by subject, decade, or region, would be tedious. Readers of a certain age will recall that by now we should have relied completely (or at least largely) on nuclear electricity, that Concorde was to be just a prelude to ubiquitous supersonic intercontinental flights, and that the Y2K glitch should have shut down all computing on January 1, 2000. But a combination of quick references to some well-known cases and brief explanations of some surprisingly little-appreciated failures provides a useful reality check—and there is no reason to assume that such misses will become less common. Moving from relatively simple pencil-and-paper forecasts to complex computerized scenarios makes it easier to perform the requisite calculations and to produce different scenarios, but it does not eliminate the inevitable perils of making assumptions. Just the opposite—more complex models combining the interactions of economic, social, technical, and environmental factors require more assumptions and open the way for greater errors.

An obvious place to start when recounting some of the now-classic forecasting failures is to look at the intellectual duel between cornucopians and catastrophists. Concerns about runaway populations surpassing available means of sustenance voiced during the 1960s could be traced to the record, and at that time still rising, rates of global population growth. For millennia, global population growth was a small fraction of a percent; it rose above 0.5 percent only during the 1770s, above 1 percent by the mid-1920s, but by the late 1950s it was close to 2 percent and still accelerating. Inevitably, many people took notice, in both professional and popular publications, and in 1960 *Science*, America's premier science periodical, succumbed to the worries about unrestrained population growth and published an absurd calculation claiming that the continuation of the historic rate of increase would result in infinitely rapid growth of the global population by November 13, 2026.[13]

This outcome—humanity growing at infinite speed—takes some imagination, but many less extreme, although still catastrophist, predictions helped to create and mobilize the modern environmental movement.[14] But there was no need to fear runaway populations: catastrophists ignored a simple fact that no form of very rapid growth can go on forever on a finite planet. Doomsday 2026 was obvious nonsense. Before the 1960s ended, global population growth reached its peak at about 2.1 percent per year, and that peak was followed by fairly rapid decline: by the year 2000 the global rate was 1.32 percent, and by 2019 it was just 1.08 percent.[15]

The halving of the relative growth rate in 50 years, and the subsequent decline of the absolute growth rate (it peaked at about 93 million a year in 1987, declined to about 80 million by 2020), changed the outlook so fundamentally that sometime during the early 2020s the world's population will cross a significant demographic milestone as half of it will be living in countries whose total fertility rate is below the replacement level.[16] This new reality immediately invites new catastrophic calculations. If this trend of falling fertility were to continue, when will the global population stop growing? And then, inevitably, when will the last *Homo sapiens* die? And a young catastrophist could speculate again about how many millions of people will die of hunger (during the 2080s?)—not because of runaway growth but because, as populations age and shrink everywhere, there will not be (even after intensive robotization) enough people of working age to feed humanity.

End-of-the-world prophecies concerning resource scarcity have not been limited to food: running out of mineral resources has been another favorite subject of catastrophic visions, and the future of crude oil, the most important energy source of 20th-century civilization, has been a favorite topic of dystopic prophecies. Forecasts of imminent peak oil extraction go back to the 1920s, but they reached new heights of existential fear-mongering during the 1990s and the first decade of the 21st century.[17] Some committed members of the peak oil cult believed that declining oil extraction would not only bring about the collapse of modern economies but that it would return humanity to a lifestyle far below its preindustrial levels, all the way to that of Paleolithic foragers—to hominins who lived in East Africa 2 million years ago.[18]

And what has actually happened? Catastrophists have always had a hard time imagining that human ingenuity can meet future food, energy, and material needs—but during the past three generations we have done so despite a tripling of the global population since 1950.

Instead of megadeaths, the share of undernourished people in low-income countries has been steadily declining, from about 40 percent during the 1960s to only about 11 percent by 2019, and average daily per capita food supply in China, the world's most populous nation, is now about 15 percent higher than in Japan.[19] Instead of desperate fertilizer shortfalls, the application of nitrogenous fertilizers has increased more than 2.5 times since 1975, and the global harvest of staple cereal crops is now about 2.2 times larger.[20] As for crude oil, its total extraction rose by two-thirds between 1995 and 2019, and at the end of that year its pre-COVID-19 price (in constant monies) was lower than it was in 2009.[21] Catastrophists are wrong, time after time.

And techno-optimists, who promise endless near-miraculous solutions, must reckon with a similarly poor record. One of the best-known (and embarrassingly well-documented) failures has been the belief in the all-encompassing power of nuclear fission. Many people appreciate that the partial success achieved by nuclear generation (it produced about 10 percent of the world's electricity in 2019, with shares at 20 percent in the US and, exceptionally, about 72 percent in France) is just a fraction of what was widely expected before 1980.[22] At that time, leading scientists and large companies not only thought that nuclear fission would eliminate all other forms of electricity generation, they also believed that the original reactors would be largely displaced by fast breeders able to produce (temporarily) more energy than they consumed. The nuclear promise went far beyond electricity generation, and some astonishingly dubious ideas were tested or expensively investigated.

Which decision was more irrational and more doomed from the very start: the pursuit of nuclear-powered flight, or natural gas production assisted by nuclear explosions? Designing a small nuclear reactor that could power submarines was one thing, making it light enough to be airborne turned out to be an insurmountable challenge—but one that was abandoned only in 1961 after spending billions of dollars on this hopeless task.[23] No plane powered by nuclear fission ever took off, but several nuclear bombs *were* detonated in the quest for expanded natural gas production. A 29-kiloton bomb (more than twice as powerful as the one dropped on Hiroshima) was detonated in

December 1967 at a depth of about 1.2 kilometers in New Mexico (code name Project Gasbuggy); in September 1969 came a 40-kiloton bomb in Colorado; in 1973 three 33-kiloton bombs, also in Colorado; and the US Atomic Energy Commission anticipated future detonations of 40–50 bombs a year.[24] There were also plans for such activities as using nuclear explosives to carve out new harbors, and using nuclear reactors to power space flight.

Little has changed half a century later: frightening prophecies and utterly unrealistic promises abound. The latest burst of intensified catastrophism has been focused on environmental degradation in general and on concerns about global climate change in particular. Journalists and activists write about climate apocalypse now, issuing final warnings. In the future, areas best suited for human habitation will shrink, large areas of the Earth are to become uninhabitable soon, climate migration will reshape America and the world, average global income will decline substantially, some prophecies claim that we might only have about a decade left to avert a global catastrophe, and in January 2020 Greta Thunberg went as far as to specify just eight years.[25]

Just a few months later, the president of the UN's General Assembly gave us 11 years to avert a complete social collapse whereupon the planet will be simultaneously burning (suffering unquenchable summer-long fires) and inundated with water (via a rapid sea-level rise). But, *nihil novi sub sole*: in 1989, another high UN official said that "governments have a 10-year window of opportunity to solve the greenhouse effect before it goes beyond human control," which means that by now we must be quite beyond the beyond, and that our very existence might be only a matter of Borgesian imagination.[26] I am convinced that we could do without this continuing flood of never-less-than-worrisome and too-often-quite-frightening predictions. How helpful is it to be told every day that the world is coming to an end in 2050 or even 2030?

Such predictably repetitive prophecies (however well-meant and however passionately presented) do not offer any practical advice about the deployment of the best possible technical solutions, about the most effective ways of legally binding global cooperation, or about tackling the difficult challenge of convincing populations of the need for significant expenditures whose benefits will not be seen for decades to come. And they are, of course, quite unnecessary according to those who argue that a "sustainable future is within our grasp," that the catastrophists have a long history of raising false alarms, who title their writings *Apocalypse Not!* and *Apocalypse Never*, and, in the starkest contradistinction to civilization's supposedly rapidly approaching final curtain, even go as far (as already noted) as seeing a not-too-distant Singularity.[27]

Why should we fear anything—be it environmental, social, or economic threats—when by 2045, or perhaps even by 2030, our understanding (or rather the intelligence unleashed by the machines we will have created) will know no bounds and hence any problem will become immeasurably less than trivial? Compared to this promise, any other recent specific and intemperate claim—from salvation through nanotechnology to fashioning new synthetic forms of life—appears trite. What will happen? An imminent near-infernal perdition, or speed-of-light godlike omnipotence?

Based on the revealed delusions of past prophecies, neither. We do not have a civilization envisioned in the early 1970s—one of worsening planetary hunger or one energized by cost-free nuclear fission—and a generation from now we will not be either at the end of our evolutionary path or have a civilization transformed by Singularity. We will still be around during the 2030s, albeit without the unimaginable benefits of speed-of-light intelligence. And we will still be trying to do the impossible, to make long-range forecasts. That is bound to bring more embarrassments and more ridiculous predictions, as well as more surprises caused by unanticipated events. Extremes are fairly easy to envisage; anticipating realities that will arise from combinations of inertial developments and unpredictable discontinuities remains an elusive quest. No amount of modeling will eliminate that, and our long-range predictions will continue to err.[28]

This is not a contradiction, not a forecast to dismiss future forecasts, just a highly probable, if not inevitable, conclusion based on the unforeseeable interplay of the inherent inertia of complex systems, with their embedded constants and long-term imperatives on one hand, and sudden discontinuities and surprises—be they technical (the rise of consumer electronics; possible breakthroughs in electricity storage) or social (the collapse of the USSR; another, much more virulent pandemic)—on the other. What makes all forecasts even harder is that now the key transformations have to unfold on enormous scales.

Inertia, scale, and mass

New departures, new solutions, and new achievements are always with us: we are a very inquisitive species with a remarkable long-term record of adaptation and with even more remarkable recent accomplishments in making the lives of most of the world's population healthier, richer, safer, and longer. Still, fundamental constraints also persist: we have changed some of them through our ingenuity, but such adjustments have their own limits. For example, we cannot eliminate the need for land, water, and nutrients in food production. As we have seen, higher yields reduced the demand for farmland and further reductions are possible if we succeed in the further closing of yield gaps (the differences between yield potential and actual harvests).

These gaps remain substantial. Even in countries that practice intensive crop cultivation (the high use of fertilizers, irrigation), the yields could go up 20–25 percent above the recent average for US corn, and 30–40 percent for Chinese rice—and, because of its still very low average productivity, they could be two to four times as high in sub-Saharan Africa.[29] In the case of high-yielding and already optimized agricultures, the resulting reduction of cropped land could be achieved with relatively small additional demands for fertilizer and irrigation. In contrast, Africa will demand substantial increases in average macronutrient applications and in the extension of irrigation. As in so many other instances, relative gains in future performance (within biological limits) should not be mistaken for an absolute decoupling of output and input variables, as long as the world's population continues to grow and as long as it demands better nutrition.

In this regard, media reports about "land-less" urban agriculture—hydroponic cultivation in high-rises—are particularly devoid of any real understanding of global food demand. Such high-input operations can produce leafy greens (lettuces, basil) and some vegetables (tomatoes, peppers) whose nutritional value is almost solely in their vitamin C content and roughage.[30] Most assuredly, hydroponic cultivation under constant lights could not be deployed to produce more than 3 billion tons of cereal and leguminous grains, whose high carbohydrate content and relatively high protein and lipid supply are required to feed nearly 8 (soon to be 10) billion people.[31]

The inertia of large, complex systems is due to their basic energetic and material demands—as well as the scale of their operations. Demands for energy and materials are constantly affected by the quest for higher efficiencies and for optimized production processes, but efficiency improvements and relative dematerialization have their physical limits, and advantages brought by new alternatives will have offsetting costs. Examples of such realities abound. Turning, once again, to two fundamental inputs, the theoretical minimum of primary energy needed to produce steel (combining blast furnace and basic oxygen furnace demands) is about 18 gigajoules per ton of hot metal, and ammonia cannot be synthesized from its elements with less than about 21 gigajoules per ton.[32]

One possible solution is to replace steel with aluminum. This lowers the mass of a specific design, but primary aluminum requires five to six times more energy to produce than primary steel, and it cannot be used in many applications that require steel's much greater strength. The most radical way to cut energy costs and the environmental impact of nitrogen fertilizers is to reduce how much is used: that option is available to affluent countries with their excessive food supply and waste—but hundreds of millions of stunted children, mostly in Africa, need to drink more milk and eat more meat, and that protein can come only from substantially increasing the amount of nitrogen they use in cropping. Just to drive this conclusion home, annual applications of fertilizers average about 160 kilograms per hectare of agricultural land in the EU and less than 20 kilograms in Ethiopia, an order-of-magnitude difference illustrating the huge development gap that is so often ignored in appraisals of global needs.[33]

And in a civilization where production of essential commodities now serves nearly 8 billion people, any departure from established practices also runs repeatedly into the constraints of scale: as we have already seen (in chapter 3), fundamental material requirements are now measured in billions and hundreds of millions of tons per year. This makes it impossible either to substitute such masses for entirely different commodities—what would take the place of more than 4 billion tons of cement or nearly 2 billion tons of steel?—or to make a rapid (years rather than decades) transition to entirely new ways of producing these essential inputs.

This inevitable inertia of mass-scale dependencies can eventually be overcome (recall that, before 1920, we had to devote a quarter of American farmland to feed crops for horses and mules), but many past examples of rapid shifts are not good guides for deriving plausible time spans for any future accomplishments. Past transitions may have been relatively fast because the magnitudes involved were comparatively small. By 1900 the world's primary energy use was roughly split between traditional biomass and fossil fuels dominated by coal, and all fossil fuels supplied an equivalent of only about 1 billion tons of coal.[34] By the year 2020, the net global supply of fossil fuels was an order of magnitude higher than the total primary energy supply in 1900, and although our technical means are now in so many ways superior, the pace of the new transition (decarbonization) has been slower than the pace of displacing traditional biomass by fossil fuels.

Even though the supply of new renewables (wind, solar, new biofuels) rose impressively, about 50-fold, during the first 20 years of the 21st century, the world's dependence on fossil carbon declined only marginally, from 87 percent to 85 percent of the total supply, and most of that small relative decline was attributable to expanded hydroelectricity generation, an old form of renewable energy.[35] Because the total energy demand was an order of magnitude lower in 1920 than it was in 2020, it was much easier to displace wood by coal in the early 20th century than it is to displace fossil fuels by new renewables (that is, to decarbonize) in the early 21st century. As a result, even a tripling or quadrupling of the recent pace of decarbonization would still leave fossil carbon dominant by 2050.

A category mistake—erroneously assigning to something a quality or action that is properly attributable only to things of another category—is behind the frequent, but deeply mistaken, conclusion that in this new, electronically enabled world everything can, and will, move much faster.[36] Information and connections do so, and so does the adoption of new personal gadgets—but existential imperatives do not belong to the category of microprocessors and mobile phones. Securing the sufficient delivery of water, growing and processing crops, feeding and slaughtering animals, producing and converting enormous quantities of primary energies, and extracting and altering raw materials to fit a myriad of uses are endeavors whose scales (required to meet the demand of billions of consumers) and infrastructures (that enable the production and distribution of these irreplaceable needs) belong to categories that are quite distinct from making a new social media profile or buying a more expensive smartphone.

Moreover, many techniques that enable these new advances are hardly new. How many people smitten by the thinness of the latest smartphone and by its ability to process information are aware that many fundamental processes that make their mass ownership possible are fairly long in the tooth? Very pure silicon is the basis of all microprocessors, including those that run all modern electronic devices—from the largest supercomputers to the smallest mobile phone—and Jan Czochralski discovered how to grow single silicon crystals in 1915. Large numbers of transistors are built into silicon, and Julius Edgar Lilienfeld patented the first field-effect transistor in 1925. And, as already detailed, integrated circuits were born in 1958–1959, and microprocessors in 1971.[37]

Most of the electricity that energizes all electronic gadgets is generated by steam turbines, machines invented by Charles A. Parsons in 1884, or by gas turbines, with the first one commercially deployed in 1938.[38] While it has been possible to replace a billion landlines by mobile phones within a generation, it will not be possible to replace terawatts of power installed in steam and gas turbines by photovoltaic cells or wind turbines within a similar time span. Mobiles, as complex as they are, are just small devices at the apex of an enormous pyramid of an industry that generates, transforms, and transmits electricity, and that requires mass-scale infrastructure to build, rebuild, and maintain.

These realities help to explain why the fundamentals of our lives will not change drastically in the coming 20–30 years, despite the near-constant flood of claims about superior innovations ranging from solar cells to lithium-ion batteries, from the 3-D printing of everything (from microparts to entire houses) to bacteria able to synthesize gasoline. Steel, cement, ammonia, and plastics will endure as the four material pillars of civilization; a major share of the world's transportation will be still energized by refined liquid fuels (automotive gasoline and diesel, aviation kerosene, and diesel and fuel oil for shipping); grain fields will be cultivated by tractors pulling plows, harrows, seeders, and fertilizer applicators and harvested by combines spilling the grains into trucks. High-rise apartments will not be printed on site by gargantuan machines, and should we soon have another pandemic then the role of the much-touted artificial intelligence will most likely be as underwhelming as it was during the 2020 SARS-CoV-2 pandemic.[39]

Ignorance, persistence, and humility

COVID-19 has provided a perfect—and costly—global reminder of our limited capacity to chart our futures, and that, too, will not (cannot) change in any dramatic way during the coming generation. The latest pandemic came after a decade that was suffused with adulatory praise of unprecedented and supposedly truly "disruptive" scientific and technical advances. Chief among them have been the anticipation of imminent deployments of the miraculous powers of artificial intelligence and neural-learning networks (Singularity-lite, one might say) and genome editing that will make it possible to engineer life forms at will.[40]

Nothing sums up better the excessive nature of these claims than the title of a 2017 bestseller, Yuval Noah Harari's *Homo Deus*.[41] And if more evidence is required, then COVID-19 exposed the emptiness of any notions of our supposed godlike capacity to control our fate: none of those much-touted capabilities was of any use in preventing the rise or controlling the diffusion of those viral RNA strands. The best we could do is what the residents of Italian towns did in the Middle Ages: stay away from others, stay inside for 40 days, isolate for *quaranta giorni*.[42] Vaccines came relatively early, but they do not cure the stricken and they do not prevent the next pandemic. And so we must pray that the next event (because there is always a next one!) will come only after decades of relatively uneventful seasonal viral epidemics, rather than in just a few years and in a much more virulent form.

COVID-19's impact in rich countries in general, and in the United States in particular, also illustrates how misplaced some of our highly touted (and very expensive) future-forming endeavors have been. Foremost among these have been the renewed steps toward manned space flight, and particularly the sci-fi-type goal of missions to Mars; trying to move toward personalized medicine (diagnosis and treatment tailored to individual patients based on their specific risk or response to a disease), with *The Economist* running a special report on this topic on March 12, 2020, just as COVID-19 began to sweep through Europe and North America filling urban hospitals with oxygen-deprived people; and being preoccupied with ever-faster connectivity, with endless hype surrounding the benefits of 5G networks.[43] How irrelevant are all of these quests while (as the cliché goes) the only remaining superpower could not provide its nurses and doctors with enough simple personal protection equipment, including such low-tech items as gloves, masks, caps, and gowns?

Consequently, the US had to pay exorbitant prices to China—the country where the brilliant architects of globalization concentrated nearly all of the manufacturing of these essential items—in order to secure airlifts of inadequate amounts of protective equipment just to prevent hospital closures in the midst of a pandemic.[44] The country that spends more than half a trillion every year on its military (more than all of its potential adversaries put together) was unprepared for an event that was absolutely certain to occur, and it did not have enough basic medical supplies: investment in domestic production worth a few hundred million dollars could have significantly reduced the economic losses of COVID-19, measured in the trillions![45]

Nor did Europe distinguish itself. Member states engaged in competition for jumbo airlifts of protective plastic from China; the vaunted absence of borders was quickly transformed into fortress arrangements; the ever-closer union failed to deliver any union-wide coordinated response; and during the first six months of the pandemic, four of the continent's five most populous nations (UK, France, Italy, and Spain) and two of its richest countries (Switzerland and Luxembourg)—whose health systems were for decades praised as paragons of excellence—recorded some of the world's highest pandemic mortalities.[46] Crises expose realities and strip away obfuscation and misdirection. The response of the affluent world to COVID-19 deserves a single ironic comment: *Homo deus* indeed!

At the same time, the rich world's reaction to COVID-19 illustrates our perennially unrealistic attitude to fundamental realities caused by forgetting even traumatic experiences. As the COVID-19 pandemic began to unfold I did not expect that this challenge would be set within proper historical perspectives (what else is to be expected in a society dominated by tweets?), and I was not surprised by references to the 1918–1919 influenza that caused the highest, albeit globally highly uncertain, number of pandemic deaths in modern history.[47] But, as I already noted in the chapter on risk, since that time we have lived through three notable (and much better understood) episodes and they have not left any deep imprint in our collective memory.

I already suggested some explanations, but others are plausible. Was the toll of more than a million deaths in 1957–1958 (in most countries taking place incrementally over 6–9 months) seen through the prism of much larger Second World War losses that were still clear in the memory of all adults? Or has our collective perception changed to such an extent that we cannot accept the fact that temporary excessive mortality will always be beyond our control? Or is it simply the fact that forgetting is an essential complement of remembering, be it on a personal or collective level, and that this, too, will not change as we will be, again and again, surprised by what should have been expected?

Persistence is as important as forgetting: despite the promises of new beginnings and bold departures, old patterns and old approaches soon resurface to set the stage for another round of failures. I ask any readers who doubt this to check sentiments during and immediately after the great financial crisis of 2007–2008—and compare them with the post-crisis experience. Who has been found responsible for this systemic near collapse of the financial order? What fundamental departures (besides enormous injections of new monies) were taken to reform questionable practices or to reduce economic inequality?[48]

Returning to the COVID-19 example, this pattern of persistence means that nobody will ever be found responsible for any of the many strategic lapses that guaranteed the mismanagement of the pandemic even before it began. Undoubtedly, some desultory hearings and a few think-tank papers will produce a list of recommendations, but those will be promptly ignored and will make no difference to deeply ingrained habits. Did the world take any resolute steps after the pandemics of 1918–1919, 1958–1959, 1968–1969, and 2009? Governments will not ensure adequate provisions of needed supplies for a future pandemic, and their response will be as inconsistent—if not as incoherent—as ever. The profits of mass-scale single-source manufacturing will not be changed for less vulnerable but more expensive decentralized production. And people will resume their constant global mingling as they return to intercontinental flights and cruises to nowhere, although it is hard to imagine a better virus incubator than a ship with 3,000 crew, and 5,000 passengers who are often mostly elderly with many pre-existing health conditions.[49]

This also means that we will need, again and again, to relearn how to reconcile ourselves to realities beyond our control. COVID-19 provides a useful reminder. The pandemic has caused the highest excessive mortality among the oldest cohorts and, as already noted, this outcome is obviously linked to our very successful efforts to extend life expectancy.[50] I, born in 1943, have been among the tens of millions of beneficiaries of this trend—but we cannot have it both ways: longer life expectancy will be accompanied by greater vulnerability. Not surprisingly, comorbidities of old age—ranging from fairly common hypertension and diabetes to less common forms of cancer and compromised immunity—have been the best predictors of excess viral mortality.[51]

Yet this will not prevent us, as it did not in 1968 or 2009, taking more steps toward prolonging life expectancy—and then fearing the likely consequences of this quest (seen, to a lesser but still substantial extent, even during seasonal influenza epidemics). Except the next time, the risk will be significantly higher because the combination of natural aging and prolongation of life will greatly increase the share of people over 65 years of age. The UN projects that share rising by about 70 percent by 2050, and in better-off countries one person in four will be older than that.[52] How will we cope in 2050 with a pandemic that might be more infectious than COVID-19, when in some countries a third of the population is in the most vulnerable category?

These realities disprove any general, automatic, embedded, unavoidable idea of progress and constant improvement that has been promoted by many techno-optimists. Neither the evolution nor the history of our species is an ever-rising arrow. There are no predictable trajectories, no definite targets. The steadily accumulating mass of our understanding and the ability to control a growing number of variables that affect our lives (ranging from food production that is sufficient to feed the world's entire population to highly effective inoculation that prevents previously dangerous infectious diseases) has lowered the overall risk of living, but it has not made many existential perils either more predictable or more manageable.

In some critical instances, our successes and our abilities to avoid the worst outcomes have been due to being prescient, vigilant, and determined to find effective fixes. Notable examples range from eliminating polio (by developing effective vaccines) to lowering the risks of commercial flying (by building more reliable airplanes and introducing better flight control measures), from reducing food pathogens (by a combination of proper food processing, refrigeration, and personal hygiene) to making childhood leukemia a largely survivable illness (by chemotherapy and stem cell transplants).[53] In other cases, we have been undoubtedly lucky: for decades we have avoided nuclear confrontation caused by an error or accident (we have experienced both on several occasions since the 1950s), not only because of built-in safeguards but also thanks to judgments that could have gone either way.[54] Again, there are no clear indications that our ability to prevent failures has been uniformly increasing.

Fukushima and the Boeing 737 MAX are, unfortunately, two perfect examples of these failures—both with large-scale and lasting consequences. Why did the Tokyo Power Company lose three reactors at its Fukushima Daiichi plant when an earthquake and tsunami struck on March 11, 2011? After all, just about 15 kilometers south of the plant, on the same Pacific shore hit by the same tsunami, its twin, Fukushima Daini, did not suffer the slightest damage. Repercussions of the Fukushima Daiichi failure have ranged from Japan being deprived of 30 percent of its electricity-generating capacity to Germany's decision to shut down all reactors by 2021—and, above all, to an even deeper public distrust of fission as a source of energy.[55]

And why has Boeing—the company that risked everything on developing the 747 in 1966, and that went on to introduce successful new families of jetliners (now up to 787s)—insisted on constantly enlarging the 737 (introduced in 1964), a dubious quest that led to two catastrophic accidents?[56] Why was the plane not grounded, either by Boeing or by the Federal Aviation Administration, immediately after the first fatal accident? Again, the consequences of these failures have been profound: first the temporary grounding of the entire 737 MAX fleet starting in March 2019, then the cessation of the plane's production and cancellation of new orders. In the long run, this will affect Boeing's ability to introduce a much-needed new design to replace its aging 757 (with all of these consequences magnified by the COVID-induced collapse of international flying).

Given the number of new designs, structures, complex processes, and interactive operations, the failures illustrated by Fukushima and the Boeing 737 MAX cannot be prevented and the coming decades will see other (and unpredictable) manifestations of this reality. The future is a replay of the past—a combination of admirable advances and (un)avoidable setbacks. But there is something new as we look ahead, that unmistakably increasing (albeit not unanimous) conviction that, of all the risks we face, global climate change is the one that needs to be tackled most urgently and effectively. And there are two fundamental reasons why this combination of speed and efficacy will be much harder to realize than is generally assumed.

Unprecedented commitments, delayed rewards

Dealing with this challenge will, for the first time in history, require a truly global, as well as a very substantial and prolonged, commitment. To conclude that we will be able to achieve decarbonization anytime soon, effectively and on the required scales, runs against all past evidence. The UN's first climate conference took place in 1992, and in the intervening decades we have had a series of global meetings and countless assessments and studies—but nearly three decades later there is still no binding international agreement to moderate the annual emissions of greenhouse gases and no prospect for its early adoption.

In order to be effective, this would have to entail nothing less than a global accord. This does not mean that 200 nations must sign on dotted lines: the combined emissions of about 50 small nations add up to less than the likely error in quantifying the emissions of just the top five greenhouse gas producers. No real progress can be achieved until at least these top five countries, now responsible for 80 percent of all emissions, agree to clear and binding commitments. But we are nowhere close to embarking on such a concerted global action.[57] Recall that the much-praised Paris accord had no specific emission-reduction targets for the world's largest emitters, and that its non-binding pledges would not mitigate anything—they would result in 50 percent higher emissions by 2050!

Moreover, any effective commitments will be expensive, they will have to last for at least two generations in order to bring the desired outcome (of much reduced, if not totally eliminated, greenhouse gas emissions), and even drastic reductions going well beyond anything that could be realistically envisaged will not show any convincing benefits for decades.[58] This raises the extraordinarily difficult problem of intergenerational justice—that is, our never-failing propensity to discount the future.[59]

We value now more than later, and we price it accordingly. An avid 30-year-old mountaineer is willing to pay some $60,000 for permits, gear, Sherpas, oxygen, and other expenses to climb Mount Everest next year. But he would demand a steep discount—reflecting such obvious intervening uncertainties as his health, the stability of future Nepali governments, the probability of major Himalayan earthquakes preventing any expeditions, and the likelihood of shutting down the access—for buying a promise to scale the mountain in 2050. This universal inclination to discount the future matters greatly when contemplating such complex and costly undertakings as pricing carbon in order to mitigate global climate change, because there would no discernible economic benefits for the generation of people that would launch the expensive quest. Because greenhouse gases remain in the atmosphere for long periods of time after they have been emitted (for CO_2, up to 200 years), even very strong mitigation efforts would not give a clear signal of success—the first significant decline of global mean surface temperature—for several decades.[60]

Obviously, a temperature rise that would continue for 25–35 years after the launching of a massive global decarbonization effort would present a major challenge for enacting and pursuing such drastic measures. But because there are currently no globally binding commitments that could see any widespread adoption of such steps within a few years, both the break-even point and the onset of measurable temperature declines advance even further into the future. A commonly used climate-economy model indicates the break-even year (when the optimal policy would begin to produce net economic benefit) for mitigation efforts launched in the early 2020s would be only around 2080.

Should average global life expectancy (about 72 years in 2020) remain the same, then the generation born near the middle of the 21st century would be the first to experience cumulative economic net benefit from climate-change mitigation policy.[61] Are the young citizens of affluent countries ready to put these distant benefits ahead of their more immediate gains? Are they willing to sustain this course for more than half a century even as the low-income countries with growing populations continue, as a matter of basic survival, to expand their reliance on fossil carbon? And are the people now in their 40s and 50s ready to join them in order to bring about rewards they will never see?

The latest pandemic has served as yet another reminder that one of the best ways to minimize the impact of increasingly global challenges is to have a set of priorities and basic measures for how to deal with them—but the pandemic, with its incoherent and non-uniform inter- and intranational measures, has also shown how difficult it would be to codify such principles and to follow such guidelines. Failures revealed during crises offer costly and convincing illustrations of our recurrent inability to get the basics right, to take care of the fundamentals. By now, readers of this book will appreciate that this (short) list must include the security of basic food, energy, and material supply, all provided with the least possible impact on the environment, and all done while realistically appraising the steps that we can take to minimize the extent of future global warming. That is a daunting prospect, and nobody can be sure that we will succeed—or that we will fail.

Being agnostic about the distant future means being honest: we have to admit the limits of our understanding, approach all planetary challenges with humility, and recognize that advances, setbacks, and failures will all continue to be a part of our evolution and that there can be no assurance of (however defined) ultimate success, no arrival at any singularity—but, as long as we use our accumulated understanding with determination and perseverance, there will also not be an early end of days. The future will emerge from our accomplishments and failures, and while we might be clever (and lucky) enough to foresee some of its forms and features, the whole remains elusive even when looking just a generation ahead.

The first draft of this closing chapter was written on May 8, 2020, the 75th anniversary of the end of the Second World War in Europe.

Let's imagine a scenario in which on that spring day in the mid-20th century a small group of people embodying all extant knowledge of the time sat down to discuss and predict the state of the world in 2020. Being aware of the latest breakthroughs in areas ranging from engineering (gas turbines, nuclear reactors, electronic computers, rockets) to life sciences (antibiotics, pesticides, herbicides, vaccines), they could correctly foresee many rising trajectories, ranging from mass automobilization and affordable intercontinental flight to electronic computing, and from rising crop yields to significant gains in life expectancy.

But they would not be able to describe the advances, complexities, and nuances of the world that we have created by our accomplishments and failures during the intervening 75 years. To stress this impossibility, just think in national terms. In 1945 Japan's wooden cities were (save for Kyoto) essentially leveled. Europe was in postwar disarray, shortly to be split by the Cold War. The USSR was victorious but at an enormous cost, and it remained under Stalin's ruthless rule. The US emerged as an unprecedented superpower, generating about half of the world's economic product. China was desperately poor and, again, on the brink of civil war. Who could have traced their specific trajectories of rise and fall (Japan), of new prosperity, new problems, new unity, and new disunity (Europe), of aggressive confidence ("We will bury you!") and demise (USSR), of blunders, defeats, wasted accomplishments, and unrealized possibilities (USA), and of suffering, the world's worst famine, slow recovery, and steep ascent to questionable heights (China)?

Nobody in 1945 could have predicted a world with more than 5 billion additional people that is also better fed than at any time in history—even as it keeps wasting an indefensibly high share of all the food it grows. Nor did anybody foresee a world that relegated a number of infectious diseases (most notably polio everywhere, and tuberculosis in affluent nations) to historical footnotes, but that cannot keep economic inequality from widening even in the richest countries; a world that is at once much cleaner and much healthier yet also more polluted in new ways (from plastic in the ocean to heavy metals in soils) and, due to the ongoing biospheric degradation, also more precarious; or a world suffused in instant and essentially free information that comes at the price of massively disseminated misinformation, lies and reprehensible claims.

A lifetime later, there is no reason to believe that we are in a better position to foresee the extent of coming technical innovations (unless, of course, you are a believer in near-imminent Singularity), the events that will shape the fortunes of nations, and the decisions (or their regrettable absence) that will determine the fate of our civilization during the next 75 years. Despite the recent preoccupation with the eventual impacts of global warming and with the need for rapid decarbonization, few uncertain outcomes will be as important in determining our future as the trajectory of the global population during the remainder of the 21st century.

Extreme forecasts offer very different futures: will the global population surpass 15 billion by 2100 (nearly twice as many as in 2020), or will it shrink to 4.8 billion, losing more than half of today's total, with China shrinking by 48 percent?[62] As expected, the medium variants of these forecasts are not that far apart (8.8 and 10.9 billion). Still, being 2 billion people apart is not an inconsequential gap, and these comparisons show how even basic population forecasts veer apart after just a single generation. All too obviously, even when forecasts go only as far as today's life expectancy in affluent countries, the implications of their extreme values shape two very different economic, social, and environmental trajectories. And as the first and second drafts of this book were written during the first and second waves of COVID-19, it is quite realistic to ask if the new pandemics we will face throughout the remainder of the 21st century (given their post-1900 frequency—1918, 1957, 1968, 2009, 2020—we can expect at least two or three such events before 2100) will be similar, much weaker, or far more virulent than the 2020 event. Living with these fundamental uncertainties remains the essence of the human condition—and it limits our ability to act with foresight.

As I noted in the opening chapter, I am not a pessimist or an optimist, I am a scientist. There is no agenda in understanding how the world *really* works.

A realistic grasp of our past, present, and uncertain future is the best foundation for approaching the unknowable expanse of time before us. While we cannot be specific, we know that the most likely prospect is a mixture of progress and setbacks, of seemingly insurmountable difficulties and near-miraculous advances. The future, as ever, is not predetermined. Its outcome depends on our actions.

Appendix: Understanding Numbers

Orders of Magnitude

Time flies, organisms grow, things change. In the world of fiction these inexorable processes and outcomes are, almost without exception, handled in qualitative terms. In fairy tales it has always been once upon a time and the protagonists are rich (princes) and poor (Cinderellas), beautiful (maidens) and ugly (ogres), bold (knights) or timid (mice). Numbers usually enter only as simple counters, as devices serving the plot, often in threes: three brothers, three wishes, three little pigs . . . Not much changes in modern fiction. Hemingway's Lady Brett Ashley is "damned good-looking" but we never learn her height, and Fitzgerald's storied Gatsby makes his first appearance just as "a man of about my age" and we never learn his age—or his real wealth. Only exact timing becomes somewhat more prominent, often in the first sentence. Zola's *Money*: "Eleven o'clock had just struck at the Bourse . . ." Faulkner's *Intruder in the Dust*: "It was just noon that Sunday morning . . ." Solzhenitsyn's *One Day in the Life of Ivan Denisovich*: "At five o'clock that morning . . ."

In contrast, today's world is full of numbers. The new fairy tales, the stories about improbable billionaires, unfailingly note the latest sums to their credit; the new tragedies, reports of the latest ferry sinking or of yet another mass homicide, always come with the number of victims. Daily counts of national and global deaths became the inescapable mark of the 2020 pandemic. Ours is a new, quantitative world, where people measure numbers of their "friends" (on Facebook), of their daily steps (by Fitbit), and of their investment prowess (by beating the NASDAQ average). This quantification is pervasive but, all too often, its quality is questionable, as the numbers range from precise and repeated measurements to sloppy assumptions and careless estimates. Unfortunately, few people who see, repeat, and use these numbers question their origins, and very few try to judge them in context. But even the best modern numbers—those that may be perfect measures of complex realities—are often elusive because they represent quantities that are either too large or too small for any intuitive comprehension.

This makes them easy subjects for misrepresentation and misuse. Even preschoolers have a mental system of magnitude representation that creates an intuitive "number sense," and this capability improves with schooling.[1] Obviously, this number system is only approximate, and it fails as the quantities rise to thousands, millions, and billions. This is where orders of magnitude come in handy. Think of them simply as the total number of digits that follow the first digit of any whole number, or the number of digits following the first digit before the decimal point. No digit follows 7 (and no additional digit comes between the first number and the decimal point in 3.5) and hence both of these numbers are of a zero order of magnitude. This is expressed on a base-10 (decadic) logarithmic scale as 10^0. Any number between 1 and 10 will be a multiple of 10^0, 10 becomes 10^1, 20 is 2×10^1. The advantages of this become quickly apparent as the numbers grow bigger. A 10-fold jump brings us to items counted in hundreds (10^2) and then to thousands (10^3), tens of thousands (10^4), hundreds of thousands (10^5), and millions (10^6).

Beyond that, we get into realms where it is easy to make order of magnitude mistakes: some rich families (business founders, owners, or lucky inheritors) now add annually to their holdings tens (10^7) or hundreds (10^8) of millions of dollars; in 2020 the world had about 2,100 billionaires (10^9 dollars), and the richest ones are now valued at more than $100 billion or 10^{11} dollars.[2] In terms of individual net worth, compared with the few dollars' worth of tattered clothes and the worn-out shoes of a destitute African migrant, the gap is thus 10 orders of magnitude.

This difference is so large that we can find no equivalent among the properties separating the two most notable classes of terrestrial animals: birds and mammals. The difference between the body masses of the smallest and the largest mammals (the Etruscan shrew at 10^0 grams and the African elephant at 10^6 grams) is "just" six orders of magnitude. The difference between the wingspan of the smallest and the largest flying birds (the bee hummingbird at 3 centimeters and the Andean condor at 320 centimeters) is only two orders of magnitude.[3] Clearly, some humans have gone far further than natural evolution ever could in separating themselves from the crowd.

And there is an even easier way to indicate orders of magnitude than by spelling out full value designations or writing them out as exponents of decadic logarithms. Because these multiples are encountered so frequently in both scientific research and in engineering practice, they were assigned specific Greek names to be used as prefixes for the first three orders of magnitude—10^1 is deka, 10^2 is hecto, 10^3 is kilo—and then for every third order: 10^6 is mega, 10^9 giga, all the way to yotta, 10^{24}, now the highest named order of magnitude. Everything—from actual numbers to specific names—is summarized in the following table:

Multiples in the International System of Units used in the text

Prefix	Abbreviation	Scientific notation
hecto	h	10^2
kilo	k	10^3
mega	M	10^6
giga	G	10^9
tera	T	10^{12}
peta	P	10^{15}
exa	E	10^{18}
zetta	Z	10^{21}
yotta	Y	10^{24}

Yet another way to illustrate the unprecedented range of magnitudes that enables the functioning of modern societies is to compare them with the range of traditional experiences. Two key examples will suffice. In preindustrial societies, the extremes of travel speeds on land differed only by a factor of two, from slow walking (4 km/hour) to horse-drawn coaches (8 km/hour) for those who could pay for a seat (often un-upholstered). In contrast, travel speeds now range over two orders of magnitude, from 4 km/hour for slow walking to 900 km/hour for jetliners.

And the most powerful prime mover (an organism or machine delivering kinetic energy) an individual could commonly control during the preindustrial era was a powerful horse at 750 watts.[4] Now hundreds of millions of people drive vehicles whose power ranges between 100 and 300 kilowatts—up to 400 times the power of a strong horse—and the pilot of a wide-body jetliner commands about 100 megawatts (equivalent to more than 130,000 strong horses) in cruising mode. These gains have been too large to be grasped directly or intuitively: understanding the modern world needs a careful attention to orders of magnitude!

References and Notes

1. Understanding Energy: Fuels and Electricity

1 We will never be able to pinpoint this event: it has been dated to as early as 3.7 and as late as 2.5 billion years ago. T. Cardona, "Thinking twice about the evolution of photosynthesis," *Open Biology* 9/3 (2019), 180246.

2 A. Herrero and E. Flores, eds., *The Cyanobacteria: Molecular Biology, Genomics and Evolution* (Wymondham: Caister Academic Press, 2008).

3 M. L. Droser and J. G. Gehling, "The advent of animals: The view from the Ediacaran," *Proceedings of the National Academy of Sciences* 112/16 (2015), pp. 4865–4870.

4 G. Bell, *The Evolution of Life* (Oxford: Oxford University Press, 2015).

5 C. Stanford, *Upright: The Evolutionary Key to Becoming Human* (Boston: Houghton Mifflin Harcourt, 2003).

6 The timing of the earliest deliberate, controlled use of fire by hominins will always remain uncertain, but the best evidence is that it was no later than about 800,000 years ago: N. Goren-Inbar et al., "Evidence of hominin control of fire at Gesher Benot Ya'aqov, Israel," *Science* 304/5671 (2004), pp. 725–727.

7 Wrangham argues that cooking was one of the most important evolutionary advances: R. Wrangham, *Catching Fire: How Cooking Made Us Human* (New York: Basic Books, 2009).

8 The domestication of numerous plant species took place independently in several regions of the Old and the New World, but the Near East produced the earliest cluster: M. Zeder, "The origins of agriculture in the Near East," *Current Anthropology* 52/Supplement 4 (2011), S221–S235.

9 Draft animals have included cattle, water buffalo, yaks, horses, mules, donkeys, camels, llamas, elephants, and (less frequently) also reindeer, sheep, goats, and dogs. Besides equines (horses, donkeys, mules), only camels, yaks, and elephants have been in common use for riding.

10 The evolution of these machines is traced in V. Smil, *Energy and Civilization: A History* (Cambridge, MA: MIT Press, 2017), pp. 146–163.

11 P. Warde, *Energy Consumption in England and Wales, 1560–2004* (Naples: Consiglio Nazionale delle Ricerche, 2007).

12 For the history of English and British coal mining see: J. U. Nef, *The Rise of the British Coal Industry* (London: G. Routledge, 1932); M. W. Flinn et al., *History of the British Coal Industry*, 5 vols (Oxford: Oxford University Press, 1984–1993).

13 R. Stuart, *Descriptive History of the Steam Engine* (London: Wittaker, Treacher and Arnot, 1829).

14 R. L. Hills, *Power from Steam: A History of the Stationary Steam Engine* (Cambridge: Cambridge University Press, 1989), p. 70; J. Kanefsky and J. Robey, "Steam engines in 18th-century Britain: A quantitative assessment," *Technology and Culture* 21 (1980), pp. 161–186.

15 These calculations are highly approximate; even if we knew the exact totals of labor force and draft animals, we would still have to make assumptions about their typical power and aggregate working hours.

16 The actual totals were less than 0.5 EJ in 1800, rising to almost 22 EJ by 1900 and to nearly 350 EJ by the year 2000, and going to 525 EJ in 2020. For a detailed historical account of global (and many national) energy transitions, see V. Smil, *Energy Transitions: Global and National Perspectives* (Santa Barbara, CA: Praeger, 2017).

17 Composite averages of historical energy efficiencies are taken from calculations I did for: Smil, *Energy and Civilization*, pp. 297–301. For overall conversion efficiencies in recent years, see Sankey diagrams of energy flows prepared for the world (https://www.iea.org/sankey) or for individual countries; and for the US, see https://flowcharts.llnl.gov/content/assets/images/energy/us/Energy_US_2019.png.

18 Data for these calculations can be found in the United Nations' *Energy Statistics Yearbook*, https://unstats.un.org/unsd/energystats/pubs/yearbook/; and in BP's *Statistical Review of World Energy*, https://www.bp.com/en/global/corporate/energy-economics/statistical-review-of-world-energy/downloads.html.

19 L. Boltzmann, *Der zweite Hauptsatz der mechanischen Wärmetheorie* (Lecture presented at the "Festive Session" of the Imperial Academy of Sciences in Vienna), May 29, 1886. See also P. Schuster, "Boltzmann and evolution: Some basic questions of biology seen with atomistic glasses," in G. Gallavotti et al., eds., *Boltzmann's Legacy* (Zurich: European Mathematical Society, 2008), pp. 1–26.

20 E. Schrödinger, *What Is Life?* (Cambridge: Cambridge University Press, 1944), p. 71.

21 A. J. Lotka, "Natural selection as a physical principle," *Proceedings of the National Academy of Sciences* 8/6 (1922), pp. 151–154.

22 H. T. Odum, *Environment, Power, and Society* (New York: Wiley Interscience, 1971), p. 27.

23 R. Ayres, "Gaps in mainstream economics: Energy, growth, and sustainability," in S. Shmelev, ed., *Green Economy Reader: Lectures in Ecological Economics and Sustainability* (Berlin: Springer, 2017), p. 40. See also R. Ayres, *Energy, Complexity and Wealth Maximization* (Cham: Springer, 2016).

24 Smil, *Energy and Civilization*, p. 1.

25 Ayres, "Gaps in mainstream economics," p. 40.

26 The history of the concept of energy is covered in revealing detail in J. Coopersmith, *Energy: The Subtle Concept* (Oxford: Oxford University Press, 2015).

27 R. S. Westfall, *Force in Newton's Physics: The Science of Dynamics in the Seventeenth Century* (New York: Elsevier, 1971).

28 C. Smith, *The Science of Energy: A Cultural History of Energy Physics in Victorian Britain* (Chicago: University of Chicago Press, 1998); D. S. L. Cardwell, *From Watt to Clausius: The Rise of Thermodynamics in the Early Industrial Age* (London: Heinemann Educational, 1971).

29 J. C. Maxwell, *Theory of Heat* (London: Longmans, Green, and Company, 1872), p. 101.

30 R. Feynman, *The Feynman Lectures on Physics* (Redwood City, CA: Addison-Wesley, 1988), vol. 4, p. 2.
31 There is no shortage of introductory books on thermodynamics, but this one still stands out: K. Sherwin, *Introduction to Thermodynamics* (Dordrecht: Springer Netherlands, 1993).
32 N. Friedman, *U.S. Submarines Since 1945: An Illustrated Design History* (Annapolis, MD: US Naval Institute, 2018).
33 Capacity (load) factor is the ratio between actual generation and the maximum output a unit is able to produce. For example, a large 5 MW wind turbine working non-stop all day would generate 120 MWh of electricity; if its actual output is only 30 MWh then its capacity factor is 25 percent. Average annual US capacity factors for 2019 were (all rounded): 21 percent for solar panels, 35 percent for wind turbines, 39 percent for hydro stations and 94 percent for nuclear stations: Table 6.07.B, "Capacity Factors for Utility Scale Generators Primarily Using Non-Fossil Fuels," https://www.eia.gov/electricity/monthly/epm_table_grapher.php?t=epmt_6_07_b. Low German solar cell capacity factor is no surprise: both Berlin and Munich have fewer sunshine hours per year than Seattle!
34 A votive candle—weighing about 50 g, with paraffin's energy density of 42 kJ/g—contains 2.1 MJ (50 × 42,000) of chemical energy, and its average power during a 15-hour burn will be nearly 40 W (much like a weak light bulb). But in both cases only a small portion of the total energy will be converted into light: less than 2 percent for a modern incandescent light, just 0.02 percent for a paraffin candle. For candle weights and burn times, see https://www.candlewarehouse.ie/shopcontent.asp?type=burn-times; for light efficacies, see https://web.archive.org/web/20120423123823/http://www.ccri.edu/physics/keefe/light.htm.
35 Calculations of basal metabolism are at: Joint FAO/WHO/UNU Expert Consultation, *Human Energy Requirements* (Rome: FAO, 2001), p. 37, http://www.fao.org/3/a-y5686e.pdf.
36 Engineering Toolbox, "Fossil and Alternative Fuels – Energy Content" (2020), https://www.engineeringtoolbox.com/fossil-fuels-energy-content-d_1298.html.

37 V. Smil, *Oil: A Beginner's Guide* (London: Oneworld, 2017); L. Maugeri, *The Age of Oil: The Mythology, History, and Future of the World's Most Controversial Resource* (Westport, CT: Praeger Publishers, 2006).

38 T. Mang, ed., *Encyclopedia of Lubricants and Lubrication* (Berlin: Springer, 2014).

39 Asphalt Institute, *The Asphalt Handbook* (Lexington, KY: Asphalt Institute, 2007).

40 International Energy Agency, *The Future of Petrochemicals* (Paris: IEA, 2018).

41 C. M. V. Thuro, *Oil Lamps: The Kerosene Era in North America* (New York: Wallace-Homestead Book Company, 1983).

42 G. Li, *World Atlas of Oil and Gas Basins* (Chichester: Wiley-Blackwell, 2011); R. Howard, *The Oil Hunters: Exploration and Espionage in the Middle East* (London: Hambledon Continuum, 2008).

43 R. F. Aguilera and M. Radetzki, *The Price of Oil* (Cambridge: Cambridge University Press, 2015); A. H. Cordesman and K. R. al-Rodhan, *The Global Oil Market: Risks and Uncertainties* (Washington, DC: CSIS Press, 2006).

44 Average performance of American cars was about 16 mpg (15 L/100 km) during the early 1930s, and it kept slowly deteriorating for four decades, reaching just 13.4 mpg (17.7 L/100 km) in 1973. The new CAFE (Corporate Average Fuel Economy) standards doubled it to 27.5 mpg (8.55 L/100 km) by 1985, but subsequent low oil prices meant no further progress until 2010. V. Smil, *Transforming the Twentieth Century* (New York: Oxford University Press, 2006), pp. 203–208.

45 Detailed statistics on energy production and consumption are available in the United Nations' *Energy Statistics Yearbook* and BP's *Statistical Review of World Energy*.

46 S. M. Ghanem, *OPEC: The Rise and Fall of an Exclusive Club* (London: Routledge, 2016); V. Smil, *Energy Food Environment* (Oxford: Oxford University Press, 1987), pp. 37–60.

47 J. Buchan, *Days of God: The Revolution in Iran and Its Consequences* (New York: Simon & Schuster, 2013); S. Maloney, *The Iranian Revolution at Forty* (Washington, DC: Brookings Institution Press, 2020).

48 Energy-intensive industries (metallurgy, chemical syntheses) were the first sectors to lower their specific energy use; success of the US CAFE standards was already noted (see note 44); and nearly all electricity generation that had previously relied on burning crude oil or a fuel oil was converted to coal or natural gas.

49 Post-1980 crude oil shares calculated from consumption data in British Petroleum, *Statistical Review of World Energy*.

50 Feynman, *The Feynman Lectures on Physics*, vol. 1, pp. 4–6.

51 These concerns now affect an increasing share of the global population—since 2007, more than half of all people live in cities, and by 2025 about 10 percent will live in megacities.

52 B. Bowers, *Lengthening the Day: A History of Lighting* (Oxford: Oxford University Press, 1988).

53 V. Smil, "Luminous efficacy," *IEEE Spectrum* (April 2019), p. 22.

54 First commercial uses of small AC electric motors came in the US during the late 1880s, and during the 1890s a small fan powered by a 125 W motor sold nearly 100,000 units: L. C. Hunter and L. Bryant, *A History of Industrial Power in the United States, 1780–1930*, vol. 3: *The Transmission of Power* (Cambridge, MA: MIT Press, 1991), p. 202.

55 S. H. Schurr, "Energy use, technological change, and productive efficiency," *Annual Review of Energy* 9 (1984), pp. 409–425.

56 Two basic designs are eccentric rotating mass vibration motors and linear vibration motors. Coin-type motors are now the thinnest (as little as 1.8 mm) available units (https://www.vibrationmotors.com/vibration-motor-product-guide/cell-phone-vibration-motor). Given the global smartphone sales—1.37 billion units in 2019 (https://www.canalys.com/newsroom/canalys-global-smartphone-market-q4-2019)—no other electric motors are now made in comparable quantities.

57 French TGV trains have two power cars whose motors have total power of 8.8–9.6 MW. In Japan's Series N700 Shinkansen, 14 out of 16 cars are motor cars with a total power of 17 MW: http://www.railway-research.org/IMG/pdf/r.1.3.3.3.pdf.

58 In luxury vehicles, the total mass of these small electric servomotors may be as much as 40 kg: G. Ombach, "Challenges and

requirements for high volume production of electric motors," SAE (2017), http://www.sae.org/events/training/symposia/emotor/presentations/2011/GrzegorzOmbach.pdf.

59 For more on electric motors in kitchen appliances, see Johnson Electric, "Custom motor drives for food processors" (2020), https://www.johnsonelectric.com/en/features/custom-motor-drives-for-food-processors.

60 Mexico City is the best example of such an extraordinary demand: water from the main source, the Cutzamala River, supplies about two-thirds of the total demand, and it must be lifted more than 1 km; with an annual delivery in excess of 300 million m^3 this represents potential energy of more than 3 PJ, equivalent to nearly 80,000 tons of diesel fuel. R. Salazar et al., "Energy and environmental costs related to water supply in Mexico City," *Water Supply* 12 (2012), pp. 768–772.

61 These are fairly small (0.25–0.5 hp; that is, about 190–370 W) motors, as even the larger blower motor has less power than the one in a small food processor (400–500 W). Forcing air is a much easier task than chopping and kneading.

62 The early history of electricity is recounted in L. Figuier, *Les nouvelles conquêtes de la science: L'électricité* (Paris: Manoir Flammarion, 1888); A. Gay and C. H. Yeaman, *Central Station Electricity Supply* (London: Whittaker & Company, 1906); M. MacLaren, *The Rise of the Electrical Industry During the Nineteenth Century* (Princeton, NJ: Princeton University Press, 1943); Smil, *Creating the Twentieth Century*, pp. 32–97.

63 Even in the US it is only slightly higher. In 2019, 27.5 percent of all US fossil fuels (roughly split between coal and natural gas, with liquid fuels accounting for a negligible share) were used to generate electricity: https://flowcharts.llnl.gov/content/assets/images/energy/us/Energy_US_2019.png.

64 International Commission on Large Dams, *World Register of Dams* (Paris: ICOLD, 2020).

65 International Atomic Energy Agency, *The Database of Nuclear Power Reactors* (Vienna: IAEA, 2020).

66 Data from British Petroleum, *Statistical Review of World Energy*.

67 Tokyo Metro, Tokyo Station Timetable (accessed 2020), https://www.tokyometro.jp/lang_en/station/tokyo/timetable/marunouchi/a/index.html.

68 A large collection of nighttime satellite images is available at https://earthobservatory.nasa.gov/images/event/79869/earth-at-night.

69 Electric Power Research Institute, *Metrics for Micro Grid: Reliability and Power Quality* (Palo Alto, CA: EPRI, 2016), http://integrat-edgrid.com/wp-content/uploads/2017/01/4-Key-Microgrid-Reliability-PQ-metrics.pdf.

70 There were no problems with electricity supply during the periods of high COVID-19 mortality, but in some cities there were temporary shortages of mortuary capacity and refrigerated trucks had to be deployed. Mortuary refrigeration is another critical sector dependent on electric motors: https://www.fiocchetti.it/en/prodotti.asp?id=7.

71 The concept recognizes that it will not be possible to eliminate all anthropogenic emissions of CO_2, but neither is there any agreement about how substantial the direct capture from the air would have to be nor any large-scale and affordable processes to do so. I will consider some of these options in the closing chapter.

72 United Nations Climate Change, "Commitments to net zero double in less than a year" (September 2020), https://unfccc.int/news/commitments-to-net-zero-double-in-less-than-a-year. See also the Climate Action Tracker (https://climateactiontracker.org/countries/).

73 The Danish Energy Agency, *Annual Energy Statistics* (2020), https://ens.dk/en/our-services/statistics-data-key-figures-and-energy-maps/annual-and-monthly-statistics.

74 German capacity and generation data can be found in: Bundesverband der Energie-und Wasserwirtschaft, *Kraftwerkspark in Deutschland* (2018), https://www.bdew.de/energie/kraftwerkspark-deutschland-gesamtfoliensatz/; VGB, Stromerzeugung 2018/2019, https://www.vgb.org/daten_stromerzeugung.html?dfid=93254.

75 Clean Line Energy, the company that planned to develop five large US transmission projects, folded in 2019, and the Plains &

Eastern Clean Line that was to become the backbone of a new US grid by 2020 (its environmental impact statement was already completed in 2014) ended with the US Department of Energy pulling out of the project; it may not be built even by 2030.

76 N. Troja and S. Law, "Let's get flexible—Pumped storage and the future of power systems," IHA website (September 2020). In 2019, Florida Power and Light announced the world's largest battery-storage, 900-MWh Manatee project to be completed in late 2021. But the largest pump hydro station (Bath County in the US) has a capacity of 24 GWh, 27 times the future FPL storage, and the global capacity of pumped hydro in 2019 was 9 TWh compared to some 7 GWh in batteries—nearly a 1,300-fold difference.

77 A single-day storage for a megacity of 20 million people would have to provide at least 300 GWh, a total more than 300 times larger than the world's largest battery storage in Florida.

78 European Commission, *Going Climate-Neutral by 2050* (Brussels: European Commission, 2020).

79 In 2019, lithium-ion batteries in the best-selling electric vehicles rated about 250 Wh/kg: G. Bower, "Tesla Model 3 2170 Energy Density Compared to Bolt, Model S1009D," InsideEVs (February 2019), https://insideevs.com/news/342679/tesla-model-3-2170-energy-density-compared-to-bolt-model-s-p100d/.

80 In January 2020, the longest scheduled flights were Newark–Singapore (9,534 km), Auckland–Doha, and Perth–London, the first one taking about 18 hours: T. Pallini, "The 10 longest routes flown by airlines in 2019," Business Insider (April 2020), https://www.businessinsider.com/top-10-longest-flight-routes-in-the-world-2020-4.

81 Bundesministerium für Wirtschaft und Energie, *Energiedaten: Gesamtausgabe* (Berlin: BWE, 2019).

82 The Energy Data and Modelling Center, *Handbook of Japan's & World Energy & Economic Statistics* (Tokyo: EDMC, 2019).

83 Consumption data from British Petroleum, *Statistical Review of World Energy*.

84 International Energy Agency, *World Energy Outlook 2020* (Paris: IEA, 2020), https://www.iea.org/reports/world-energy-outlook-2020.

85 V. Smil, "What we need to know about the pace of decarbonization," *Substantia* 3/2, supplement 1 (2019), pp. 13–28; V. Smil, "Energy (r)evolutions take time," *World Energy* 44 (2019), pp. 10–14. For a different perspective, see Energy Transitions Commission, *Mission Possible: Reaching Net-Zero Carbon Emissions from Harder-to-Abate Sectors by Mid-Century* (2018), http://www.energy-transitions.org/sites/default/files/ETC_MissionPossible_FullReport.pdf.

2. Understanding Food Production: Eating Fossil Fuels

1 B. L. Pobiner, "New actualistic data on the ecology and energetics of hominin scavenging opportunities," *Journal of Human Evolution* 80 (2015), pp. 1–16; R. J. Blumenschine and J. A. Cavallo, "Scavenging and human evolution," *Scientific American* 267/4 (1992), pp. 90–95.

2 V. Smil, *Energy and Civilization: A History* (Cambridge, MA: MIT Press 2018), pp. 28–40.

3 K. W. Butzer, *Early Hydraulic Civilization in Egypt* (Chicago: University of Chicago Press, 1976); K. W. Butzer, "Long-term Nile flood variation and political discontinuities in Pharaonic Egypt," in J. D. Clark and S. A. Brandt, eds., *From Hunters to Farmers* (Berkeley: University of California Press 1984), pp. 102–112.

4 FAO, *The State of Food Security and Nutrition in the World* (Rome: FAO, 2020), http://www.fao.org/3/ca9692en/CA9692EN.pdf.

5 Wavelengths that are largely absorbed are 450–490 nm for the blue and 635–700 nm for the red part of the spectrum; green (520–560 nm) is largely reflected, hence the vegetation's dominant color.

6 Total annual productivity of terrestrial (forests, grasslands, crops) and oceanic (mostly phytoplankton) photosynthesis is roughly the same, but unlike land plants, phytoplankton is very short-lived, persisting for just a few days.

7 Detailed accounts of 19th-century American cropping practices are compiled in L. Rogin, *The Introduction of Farm Machinery* (Berkeley: University of California Press, 1931). The time budget

for 1800 is based on practices prevailing between 1790 and 1820, detailed on p. 234.

8 Calculations based on Rogin's data for wheat cultivation in North Dakota's Richland county in 1893, p. 218.

9 Smil, *Energy and Civilization*, p. 111.

10 For the average size of US farms between 1850 and 1940, see US Department of Agriculture, *U.S. Census of Agriculture: 1940*, p. 68. For the size of Kansas farms: Kansas Department of Agriculture, Kansas Farm Facts (2019), https://agriculture.ks.gov/about-kda/kansas-agriculture.

11 For pictures and technical specifications of large tractors, see John Deere's website at https://www.deere.com/en/agriculture/.

12 My calculations are based on 2020 crop budgets for non-irrigated Kansas wheat and on typical work rate estimates: Kansas State University, *2020 Farm Management Guides for Non-Irrigated Crops*, https://www.agmanager.info/farm-mgmt-guides/2020-farm-management-guides-non-irrigated-crops; B. Battel and D. Stein, *Custom Machine and Work Rate Estimates* (2018), https://www.canr.msu.edu/field_crops/uploads/files/2018 percent20Custom percent20Machine percent20Work percent20Rates.pdf.

13 Quantification of these indirect energy uses requires many inevitable assumptions and approximations, and hence it can never be as accurate as the monitoring of direct fuel consumption.

14 For example, European applications of glyphosate, the world's most widely used herbicide, average just 100–300 g of active ingredient per hectare: C. Antier, "Glyphosate use in the European agricultural sector and a framework for its further monitoring," *Sustainability* 12 (2020), p. 5682.

15 V. Gowariker et al., *The Fertilizer Encyclopedia* (Chichester: John Wiley, 2009); H. F. Reetz, *Fertilizers and Their Efficient Use* (Paris: International Fertilizer Association, 2016).

16 But the crop that has received by far the highest nitrogen applications is Japan's green tea. Its dry leaves contain 5–6 percent nitrogen; plantations get commonly more than 500 kg N/ha and as much as 1 t N/ha: K. Oh et al., "Environmental problems from

tea cultivation in Japan and a control measure using calcium cyanamide," *Pedosphere* 16/6 (2006), pp. 770–777.

17 G. J. Leigh, ed., *Nitrogen Fixation at the Millennium* (Amsterdam: Elsevier, 2002); T. Ohyama, ed., *Advances in Biology and Ecology of Nitrogen Fixation* (IntechOpen, 2014), https://www.intechopen.com/books/advances-in-biology-and-ecology-of-nitrogen-fixation.

18 Sustainable Agriculture Research and Education, *Managing Cover Crops Profitably* (College Park, MD: SARE, 2012).

19 Émile Zola, *The Fat and the Thin*, https://www.gutenberg.org/files/5744/5744-h/5744-h.htm.

20 For the history of ammonia synthesis see: V. Smil, *Enriching the Earth: Fritz Haber, Carl Bosch, and the Transformation of World Food Production* (Cambridge, MA: MIT Press, 2001); D. Stoltzenberg, *Fritz Haber: Chemist, Nobel Laureate, German, Jew* (Philadelphia, PA: Chemical Heritage Press, 2004).

21 N. R. Borlaug, *The Green Revolution Revisited and The Road Ahead*, Nobel Prize Lecture 1970, https://assets.nobelprize.org/uploads/2018/06/borlaug-lecture.pdf; M. S. Swaminathan, *50 Years of Green Revolution: An Anthology of Research Papers* (Singapore: World Scientific Publishing, 2017).

22 G. Piringer and L. J. Steinberg, "Reevaluation of energy use in wheat production in the United States," *Journal of Industrial Ecology* 10/1–2 (2006), pp. 149–167; C. G. Sørensen et al., "Energy inputs and GHG emissions of tillage systems, *Biosystems Engineering* 120 (2014), pp. 2–14; W. M. J. Achten and K. van Acker, "EU-average impacts of wheat production: A meta-analysis of life cycle assessments," *Journal of Industrial Ecology* 20/1 (2015), pp. 132–144; B. Degerli et al., "Assessment of the energy and exergy efficiencies of farm to fork grain cultivation and bread making processes in Turkey and Germany," *Energy* 93 (2015), pp. 421–434.

23 Diesel fuel is used by all large farm machinery (tractors, combines, trucks, irrigation pumps) as well as in long-distance bulk transportation of crops (freight trains pulled by diesel locomotives, barges, ships). Small tractors and pickup trucks run on gasoline, and propane is used for grain drying.

24 This is just a bit less voluminous than the US cup used to measure cooking ingredients: it is exactly 236.59 mL.

25 N. Myhrvold and F. Migoya, *Modernist Bread* (Bellevue, WA: The Cooking Lab, 2017), vol. 3, p. 63.

26 Bakerpedia, "Extraction rate," https://bakerpedia.com/processes/extraction-rate/.

27 Carbon Trust, *Industrial Energy Efficiency Accelerator: Guide to the Industrial Bakery Sector* (London: Carbon Trust, 2009); K. Andersson and T. Ohlsson, "Life cycle assessment of bread produced on different scales," *International Journal of Life Cycle Assessment* 4 (1999), pp. 25–40.

28 For details on broiler CAFOs, see V. Smil, *Should We Eat Meat?* (Chichester: Wiley-Blackwell, 2013), pp. 118–127, 139–149.

29 US Department of Agriculture, *Agricultural Statistics* (2019), USDA Table 1–75, https://www.nass.usda.gov/Publications/Ag_Statistics/2019/2019_complete_publication.pdf.

30 National Chicken Council, "U.S. Broiler Performance" (2020), https://www.nationalchickencouncil.org/about-the-industry/statistics/u-s-broiler-performance/.

31 For comparisons of live, carcass, and edible weights for domestic meat animals, see Smil, *Should We Eat Meat?*, pp. 109–110.

32 V. P. da Silva et al., "Variability in environmental impacts of Brazilian soybean according to crop production and transport scenarios," *Journal of Environmental Management* 91/9 (2010), pp. 1831–1839.

33 M. Ranjaniemi and J. Ahokas, "A case study of energy consumption measurement system in broiler production," *Agronomy Research Biosystem Engineering* Special Issue 1 (2012), pp. 195–204; M. C. Mattioli et al., "Energy analysis of broiler chicken production system with darkhouse installation," *Revista Brasileira de Engenharia Agrícola e Ambienta* 22 (2018), pp. 648–652.

34 US Bureau of Labor Statistics, "Average Retail Food and Energy Prices, U.S. and Midwest Region" (accessed 2020), https://www.bls.gov/regions/mid-atlantic/data/averageretailfoodandenergyprices_usandmidwest_table.htm; FranceAgriMer, "Poulet" (accessed 2020), https://rnm.franceagrimer.fr/prix?POULET.

35 R. Mehta, "History of tomato (poor man's apple)," *IOSR Journal of Humanities and Social Science* 22/8 (2017), pp. 31–34.

36 A tomato contains about 20 mg of vitamin C per 100 g; the recommended daily dietary intake of vitamin C is 60 mg for adults.

37 D. P. Neira et al, "Energy use and carbon footprint of the tomato production in heated multi-tunnel greenhouses in Almeria within an exporting agri-food system context," *Science of the Total Environment* 628 (2018), pp. 1627–1636.

38 Almería tomato crops receive 1,000–1,500 kg N/ha in a year, while an average crop of US grain corn gets 150 kg N/ha: US Department of Agriculture, *Fertilizer Use and Price* (2020), table 10, https://www.ers.usda.gov/data-products/fertilizer-use-and-price.aspx.

39 "Spain: Almeria already exports 80 percent of the fruit and veg it produces," Fresh Plaza (2018), https://www.freshplaza.com/article/9054436/spain-almeria-already-exports-80-of-the-fruit-and-veg-it-produces/.

40 Typical fuel consumption of European long-distance trucks is 30 L/100 km or 11 MJ/km: International Council of Clean Transportation, *Fuel Consumption Testing of Tractor-Trailers in the European Union and the United States* (May 2018).

41 Industrial-scale fishing now takes place in more than 55 percent of the world's oceans, encompassing an area that is more than four times as large as that dedicated to global agriculture: D. A. Kroodsma et al., "Tracking the global footprint of fisheries," *Science* 359/6378 (2018), pp. 904–908. Illegally fishing ships turn off their transponders, but the locations of thousands of legally operating fishing vessels (orange markers) can be seen in real time at https://www.marinetraffic.com.

42 R. W. R. Parker and P. H. Tyedmers, "Fuel consumption of global fishing fleets: Current understanding and knowledge gaps," *Fish and Fisheries* 16/4 (2015), pp. 684–696.

43 The highest energy cost is for crustaceans (shrimps and lobster) caught by destructive bottom trawls in Europe, with maxima up to 17.3 L/kg of catch.

44 D. A. Davis, *Feed and Feeding Practices in Aquaculture* (Sawston: Woodhead Publishing, 2015); A. G. J. Tacon et al., "Aquaculture

feeds: addressing the long-term sustainability of the sector," in *Farming the Waters for People and Food* (Rome: FAO, 2010), pp. 193–231.

45 S. Gingrich et al., "Agroecosystem energy transitions in the old and new worlds: trajectories and determinants at the regional scale," *Regional Environmental Change* 19 (2018), pp. 1089–1101; E. Aguilera et al., *Embodied Energy in Agricultural Inputs: Incorporating a Historical Perspective* (Seville: Pablo de Olavide University, 2015); J. Woods et al., "Energy and the food system," *Philosophical Transactions of the Royal Society B: Biological Sciences* 365 (2010), pp. 2991–3006.

46 V. Smil, *Growth: From Microorganisms to Megacities* (Cambridge, MA: MIT Press, 2019), p. 311.

47 S. Hicks, "Energy for growing and harvesting crops is a large component of farm operating costs," Today in Energy (October 17, 2014), https://www.eia.gov/todayinenergy/detail.php?id=18431.

48 P. Canning et al., *Energy Use in the U.S. Food System* (Washington, DC: USDA, 2010).

49 Farm consolidation has been steadily progressing: J. M. MacDonald et al., "Three Decades of Consolidation in U.S. Agriculture," USDA Economic Information Bulletin 189 (March 2018). Food imports as a share of total consumption have been rising even in many nations that are large net food exporters (USA, Canada, Australia, France), mainly because of higher demand for fresh fruits, vegetables, and seafood. Since 2010, the share of Americans' budget for food away from home has been surpassing the share for food at home: M. J. Saksena et al., *America's Eating Habits: Food Away From Home* (Washington, DC: USDA, 2018).

50 S. Lebergott, "Labor force and Employment, 1800–1960," in D. S. Brady, ed., *Output, Employment, and Productivity in the United States After 1800* (Cambridge, MA: NBER, 1966), pp. 117–204.

51 Smil, *Growth*, pp. 122–124.

52 For the nitrogen content of many kinds of organic wastes, see Smil, *Enriching the Earth*, appendix B, pp. 234–236. For nitrogen content of fertilizers, see *Yara Fertilizer Industry Handbook 2018*, https://www.yara.com/siteassets/investors/057-reports-and-pre

sentations/other/2018/fertilizer-industry-handbook-2018-with-notes.pdf/.

53 I calculated global nitrogen flows in crop production for the mid-1990s (V. Smil, "Nitrogen in crop production: An account of global flows," *Global Biogeochemical Cycles* 13 (1999), pp. 647–662) and used the latest available data on harvests and animal counts to prepare an updated version for 2020.

54 C. M. Long et al., "Use of manure nutrients from concentrated animal feeding operations," *Journal of Great Lakes Research* 44 (2018), pp. 245–252.

55 X. Ji et al., "Antibiotic resistance gene abundances associated with antibiotics and heavy metals in animal manures and agricultural soils adjacent to feedlots in Shanghai; China," *Journal of Hazardous Materials* 235–236 (2012), pp. 178–185.

56 FAO, *Nitrogen Inputs to Agricultural Soils from Livestock Manure: New Statistics* (Rome: FAO, 2018).

57 Volatilized ammonia is also a threat to human health: its reaction with acidic compounds in the atmosphere forms fine particles that cause lung diseases, and ammonia deposited to land or waters may cause excessive nitrogen loads: S. G. Sommer et al., "New emission factors for calculation of ammonia volatilization from European livestock manure management systems," *Frontiers in Sustainable Food Systems* 3 (November 2019).

58 For typical ranges of biofixation by leguminous cover crops, see Smil, *Enriching the Earth*, appendix C, p. 237. Average nitrogen applications to major US crops are available at: US Department of Agriculture, *Fertilizer Use and Price*, https://www.ers.usda.gov/data-products/fertilizer-use-and-price.aspx. The declining supply of pulses is documented at http://www.fao.org/faostat/en/#data/FBS.

59 Recent average global yields have been about 4.6 t/ha for rice and 3.5 t/ha for wheat—and 2.7 t/ha for soybeans and just 1.1 t/ha for lentils. Yield gaps are far larger in China: 7 t/ha for rice and 5.4 t/ha for wheat, compared to 1.8 t/ha for soybeans and 3.7 t/ha for peanuts (China's other favorite pulse). Data from: http://www.fao.org/faostat/en/#data.

60 Double-cropping means either growing the same crop in succession during the same year (common with rice in China), or following a leguminous crop by a grain crop (for example the peanut/wheat rotation common on the North China Plain).

61 S.-J. Jeong et al., "Effects of double cropping on summer climate of the North China Plain and neighbouring regions," *Nature Climate Change* 4/7 (2014), pp. 615–619; C. Yan et al., "Plastic-film mulch in Chinese agriculture: Importance and problem," *World Agriculture* 4/2 (2014), pp. 32–36.

62 For the numbers of people supported per unit area of cropland, see Smil, *Enriching the Earth*.

63 Average daily intake for all Americans older than two years is about 2,100 kcal, while the average per capita supply is 3,600 kcal, a difference of more than 70 percent! Similar gaps apply to most EU countries, and among affluent nations only Japan's supply is much closer to actual consumption (about 2,700 vs. 2,000 kcal/day).

64 FAO, *Global Initiative on Food Loss and Waste Reduction* (Rome: FAO, 2014).

65 WRAP, *Household food waste: Restated data for 2007–2015* (2018).

66 USDA, "Food Availability (Per Capita) Data System," https://www.ers.usda.gov/data-products/food-availability-per-capita-data-system/.

67 China's average daily food supply is now about 3,200 kcal/capita compared to the Japanese mean of about 2,700 kcal/capita. On China's food waste, see H. Liu, "Food wasted in China could feed 30-50 million: Report," *China Daily* (March 2018).

68 The average US family now spends only 9.7 percent of its disposable income on food; EU averages range from 7.8 percent in the UK to 27.8 percent in Romania: Eurostat, "How much are households spending on food?" (2019).

69 C. B. Stanford and H. T. Bunn, eds., *Meat-Eating and Human Evolution* (New York: Oxford University Press, 2001); Smil, *Should We Eat Meat?*

70 For carnivory among common chimpanzees, see C. Boesch, "Chimpanzees—red colobus: A predator-prey system," *Animal*

Behaviour 47 (1994), pp. 1135–1148; C. B. Stanford, *The Hunting Apes: Meat Eating and the Origins of Human Behavior* (Princeton: Princeton University Press, 1999). For carnivory among bonobos, see G. Hohmann and B. Fruth, "Capture and meat eating by bonobos at Lui Kotale, Salonga National Park, Democratic Republic of Congo," *Folia Primatologica* 79/2 (2008), pp. 103–110.

71 Detailed Japanese historical statistics document this trend. In 1900, 17-year-old students averaged 157.9 cm; by 1939 the mean was 162.5 cm (increment of 1.1 mm/year); wartime and postwar food shortages lowered it to 160.6 cm by 1948; but by the year 2000 better nutrition boosted it to 170.8 cm (increment of about 0.2 mm/year): Statistics Bureau, Japan, *Historical Statistics of Japan* (Tokyo: Statistics Bureau, 1996).

72 Z. Hrynowski, "What percentage of Americans are vegetarians?" Gallup (September 2019), https://news.gallup.com/poll/267074/percentage-americans-vegetarian.aspx.

73 Annual per capita meat supply (carcass weight) is available from: http://www.fao.org/faostat/en/#data/FBS.

74 For details on changing French meat-eating habits, see C. Duchène et al., *La consommation de viande en France* (Paris: CIV, 2017).

75 The European Union now uses about 60 percent of its total grain (wheat, corn, barley, oats, and rye) production for feeding: USDA, *Grain and Feed Annual 2020*.

76 Based on per capita meat supply (carcass weight) averages: http://www.fao.org/faostat/en/#data/FBS.

77 L. Lassaletta et al., "50 year trends in nitrogen use efficiency of world cropping systems: the relationship between yield and nitrogen input to cropland," *Environmental Research Letters* 9 (2014), 105011.

78 J. Guo et al., "The rice production practices of high yield and high nitrogen use efficiency in Jiangsu," *Nature Scientific Reports* 7 (2016), article 2101.

79 The first demonstration electric tractor prototype built by John Deere, the world's leading tractor company, has no batteries: it is powered by a kilometer-long cable carried on an attached reel, an interesting but hardly a convenient universal solution:

https://enrg.io/john-deere-electric-tractor-everything-you-need-to-know/.

80 M. Rosenblueth et al., "Nitrogen fixation in cereals," *Frontiers in Microbiology* 9 (2018), p. 1794; D. Dent and E. Cocking, "Establishing symbiotic nitrogen fixation in cereals and other non-legume crops: The Greener Nitrogen Revolution," *Agriculture & Food Security* 6 (2017), p. 7.

81 H. T. Odum, *Environment, Power, and Society* (New York: Wiley-Interscience, 1971), pp. 115–116.

3. Understanding Our Material World: The Four Pillars of Modern Civilization

1 The first commercial product using transistors was a Sony radio in 1954; the first microprocessor was Intel's 4004 in 1971; the first widely used personal computer was the Apple II introduced in 1977, followed by the IBM PC in 1981, and IBM released the first smartphone in 1992.

2 P. Van Zant, *Microchip Fabrication: A Practical Guide to Semiconductor Processing* (New York: McGraw-Hill Education, 2014). For energy costs, see M. Schmidt et al., "Life cycle assessment of silicon wafer processing for microelectronic chips and solar cells," *International Journal of Life Cycle Assessment* 17 (2012), pp. 126–144.

3 Semiconductor and Materials International, "Silicon shipment statistics" (2020), https://www.semi.org/en/products-services/market-data/materials/si-shipment-statistics.

4 V. Smil, *Making the Modern World: Materials and Dematerialization* (Chichester: John Wiley, 2014); Smil, "What we need to know about the pace of decarbonization." For more on the energy cost of materials, see T. G. Gutowski et al., "The energy required to produce materials: constraints on energy-intensity improvements, parameters of demand," *Philosophical Transactions of the Royal Society A* 371 (2013), 20120003.

5 Annual totals of national and global output of all commercially important metals and non-metallic minerals are available in regular

updates published by the US Geological Survey. The latest edition is: US Geological Survey, *Mineral Commodity Summaries 2020*, https://pubs.usgs.gov/periodicals/mcs2020/mcs2020.pdf.

6 J. P. Morgan, *Mountains and Molehills: Achievements and Distractions on the Road to Decarbonization* (New York: J. P. Morgan Private Bank, 2019).

7 These are my approximate calculations, based on annual output of 1.8 Gt of steel, 4.5 Gt of cement, 150 Mt of NH_3, and 370 Mt of plastics.

8 Smil, "What we need to know about the pace of decarbonization." For an optimistic view of decarbonization possibilities of harder-to-abate sectors, see Energy Transitions Commission, *Mission Possible*.

9 M. Appl, *Ammonia: Principles & Industrial Practice* (Weinheim: Wiley-VCH, 1999); Smil, *Enriching the Earth*.

10 Science History Institute, "Roy J. Plunkett," https://www.sciencehistory.org/historical-profile/roy-j-plunkett.

11 For details, see V. Smil, *Grand Transitions: How the Modern World Was Made* (New York: Oxford University Press, 2021).

12 For the history of global land use changes, see HYDE, *History Database of the Global Environment* (2010), http://themasites.pbl.nl/en/themasites/hyde/index.html.

13 Florida and North Carolina still produce more than 75 percent of American phosphate rock, which now accounts for about 10 percent of the global output: USGS, "Phosphate rock" (2020), https://pubs.usgs.gov/periodicals/mcs2020/mcs2020-phosphate.pdf.

14 Smil, *Enriching the Earth*, pp. 39–48.

15 W. Crookes, *The Wheat Problem* (London: John Murray, 1899), pp. 45–46.

16 For the precursors of Haber's discovery and for detailed descriptions of his laboratory experiments, see Smil, *Enriching the Earth*, pp. 61–80.

17 For the life and work of Carl Bosch, see K. Holdermann, *Im Banne der Chemie: Carl Bosch Leben und Werk* (Düsseldorf: Econ-Verlag, 1954).

18 At that time, the share of inorganic nitrogen fertilizers in China's agricultural supply was no more than 2 percent: Smil, *Enriching the Earth*, p. 250.

19 V. Pattabathula and J. Richardson, "Introduction to ammonia production," *CEP* (September 2016), pp. 69–75; T. Brown, "Ammonia technology portfolio: optimize for energy efficiency and carbon efficiency," Ammonia Industry (2018); V. S. Marakatti and E. M. Giagneaux, "Recent advances in heterogeneous catalysis for ammonia synthesis," *ChemCatChem* (2020).

20 V. Smil, *China's Past, China's Future: Energy, Food, Environment* (London: RoutledgeCurzon, 2004), pp. 72–86.

21 For details on M. W. Kellogg's ammonia process, see Smil, *Enriching the Earth*, pp. 122–130.

22 FAO, http://www.fao.org/faostat/en/#search/Food%20supply%20kcal%2Fcapita%2Fday.

23 L. Ma et al., "Modeling nutrient flows in the food chain of China," *Journal of Environmental Quality* 39/4 (2010), pp. 1279–1289. India's share is also that high: H. Pathak et al., "Nitrogen, phosphorus, and potassium in Indian agriculture," *Nutrient Cycling in Agroecosystems* 86 (2010), pp. 287–299.

24 I am always amused when I see yet another list of the most important (or the greatest) modern inventions containing computers or nuclear reactors or transistors or automobiles . . . and always missing ammonia synthesis!

25 Annual per capita meat consumption (carcass weight) is a good indicator of these differences: the recent averages have been about 120 kg in the US, 60 kg in China and just 4 kg in India: http://www.fao.org/faostat/en/#data/FBS.

26 The stain-breaking power of ammonia makes it a favorite ingredient. Windex, the most common North American window-cleaning liquid, contains 5 percent NH_3.

27 J. Sawyer, "Understanding anhydrous ammonia application in soil" (2019), https://crops.extension.iastate.edu/cropnews/2019/03/understanding-anhydrous-ammonia-application-soil.

28 *Yara Fertilizer Industry Handbook*.

29 East and South Asia (dominated, respectively, by China and India) now consumes just over 60 percent of all urea: Nutrien, *Fact Book 2019*, https://www.nutrien.com/sites/default/files/uploads/2019-05/Nutrien%20Fact%20Book%202019.pdf.

30 The global average of the absorption of applied nitrogen by crops (efficiency of fertilizer use) decreased between 1961 and 1980 (from 68 percent to 45 percent), and it has since stabilized at about 47 percent: L. Lassaletta et al., "50 year trends in nitrogen use efficiency of world cropping systems: the relationship between yield and nitrogen input to cropland," *Environmental Research Letters* 9 (2014), 105011.

31 J. E. Addicott, *The Precision Farming Revolution: Global Drivers of Local Agricultural Methods* (London: Palgrave Macmillan, 2020).

32 Calculated from data at: http://www.fao.org/faostat/en/#data/RFN.

33 Europe now applies 3.5 times more nitrogen per hectare of cropland than Africa, and the differences among the most intensively fertilized land in the EU and the poorest croplands of the sub-Saharan Africa are more than tenfold: http://www.fao.org/faostat/en/#data/RFN.

34 Some common polymerization reactions—processes of converting simpler (monomer) molecules into longer-chained three-dimensional networks—require an only marginally larger mass of the initial input: 1.03 units of ethylene are needed to make 1 unit of low-density polyethylene (whose most common use is in plastic bags), and the same ratio applies to converting vinyl chloride into polyvinyl chloride (PVC, common in health-care products). P. Sharpe, "Making plastics: from monomer to polymer," *CEP* (September 2015).

35 M. W. Ryberg et al., *Mapping of Global Plastics Value Chain and Plastics Losses to the Environment* (Paris: UNEP, 2018).

36 The Engineering Toolbox, "Young's Modulus—Tensile and Yield Strength for Common Materials" (2020), https://www.engineeringtoolbox.com/young-modulus-d_417.html.

37 The Boeing 787 was the first airliner built predominantly of composite materials: by volume they make up 89 percent of the plane and by weight 50 percent, with 20 percent being aluminum, 15 percent titanium, and 10 percent steel: J. Hale, "Boeing 787 from the ground up," *Boeing AERO* 24 (2006), pp. 16–23.

38 W. E. Bijker, *Of Bicycles, Bakelites, and Bulbs: Toward a Theory of Sociotechnical Change* (Cambridge, MA: The MIT Press, 1995).
39 S. Mossman, ed., *Early Plastics: Perspectives, 1850–1950* (London: Science Museum, 1997); S. Fenichell, *Plastic: The Making of a Synthetic Century* (New York: HarperBusiness, 1996); R. Marchelli, *The Civilization of Plastics: Evolution of an Industry Which has Changed the World* (Pont Canavese: Sandretto Museum, 1996).
40 N. A. Barber, *Polyethylene Terephthalate: Uses, Properties and Degradation* (Haupaugge, NY: Nova Science Publishers, 2017).
41 P. A. Ndiaye, *Nylon and Bombs: DuPont and the March of Modern America* (Baltimore, MD: Johns Hopkins University Press, 2006).
42 R. Geyer et al., "Production, use, and fate of all plastic ever made," *Science Advances* 3 (2017), e1700782.
43 And not just all kinds of small plastic items: floors, room dividers, ceiling tiles, doors, and window frames may be plastic as well.
44 Here is a comprehensive review of American PPE shortages: S. Gondi et al., "Personal protective equipment needs in the USA during the COVID-19 pandemic," *The Lancet* 390 (2020), e90–e91. And here is just one of scores of media reports: Z. Schlanger, "Begging for Thermometers, Body Bags, and Gowns: U.S. Health Care Workers Are Dangerously Ill-Equipped to Fight COVID-19," *Time* (April 20, 2020). For a global perspective, see World Health Organization, "Shortage of personal protective equipment endangering health workers worldwide" (3 March 2020).
45 C. E. Wilkes and M. T. Berard, *PVC Handbook* (Cincinnati, OH: Hanser, 2005).
46 M. Eriksen et al., "Plastic pollution in the world's oceans: More than 5 trillion plastic pieces weighing over 250,000 tons afloat at sea," *PLoS ONE* 9/12 (2014) e111913. And here is the explanation of why most of them are not plastic: G. Suaria et al., "Microfibers in oceanic surface waters: A global characterization," *Science Advances* 6/23 (2020).
47 Basic graphs and tables summarizing steel and cast iron classification are available at: https://www.mah.se/upload/_upload/steel%20 and %20cast%20iron.pdf.

48 For the long history of pig iron, see V. Smil, *Still the Iron Age: Iron and Steel in the Modern World* (Amsterdam: Elsevier, 2016), pp. 19–31.

49 For details on premodern ways of steelmaking in Japan, China, India, and Europe, see Smil, *Still the Iron Age*, pp. 12–17.

50 Compressive strengths of granite and steel are up to 250 million pascals (MPa) but granite's tensile strength is no higher than 25 MPa compared to 350–750 MPa for construction steels: Cambridge University Engineering Department, *Materials Data Book* (2003), http://www-mdp.eng.cam.ac.uk/web/library/enginfo/cued databooks/materials.pdf.

51 For the most detailed treatment, see J. E. Bringas, ed., *Handbook of Comparative World Steel Standards* (West Conshohocken, PA: ASTM International, 2004).

52 M. Cobb, *The History of Stainless Steel* (Materials Park, OH: ASM International, 2010).

53 Council on Tall Buildings and Human Habitat, "Burj Khalifa" (2020), http://www.skyscrapercenter.com/building/burj-khalifa/3.

54 The Forth Bridges, "Three bridges spanning three centuries" (2020), https://www.theforthbridges.org/.

55 D. MacDonald and I. Nadel, *Golden Gate Bridge: History and Design of an Icon* (San Francisco: Chronicle Books, 2008).

56 "Introduction of Akashi-Kaikyō Bridge," Bridge World (2005), https://www.jb-honshi.co.jp/english/bridgeworld/bridge.html.

57 J. G. Speight, *Handbook of Offshore Oil and Gas Operations* (Amsterdam: Elsevier, 2011).

58 Smil, *Making the Modern World*, p. 61.

59 World Steel Association, "Steel in Automotive" (2020), https://www.worldsteel.org/steel-by-topic/steel-markets/automotive.html.

60 International Association of Motor Vehicle Manufacturers, "Production Statistics" (2020), http://www.oica.net/production-statis tics/.

61 Nippon Steel Corporation, "Rails" (2019), https://www.nippon steel.com/product/catalog_download/pdf/K003en.pdf.

62 For the history of container ships, see V. Smil, *Prime Movers of Globalization* (Cambridge, MA: MIT Press, 2010), pp. 180–194.

63 U.S. Bureau of Transportation Statistics, "U.S. oil and gas pipelinemileage" (2020), https://www.bts.gov/content/us-oil-and-gas-pipeline-mileage.

64 Main battle tanks are the heaviest steel weapons deployed on a large scale by modern armies: the largest version of the US M1 Abrams tank (nearly all steel) weighs 66.8 tons.

65 D. Alfè et al., "Temperature and composition of the Earth's core," *Contemporary Physics* 48/2 (2007), pp. 63–68.

66 Sandatlas, "Composition of the crust" (2020), https://www.sandatlas.org/composition-of-the-earths-crust/.

67 US Geological Survey, "Iron ore" (2020), https://pubs.usgs.gov/periodicals/mcs2020/mcs2020-iron-ore.pdf.

68 A. T. Jones, *Electric Arc Furnace Steelmaking* (Washington, DC: American Iron and Steel Institute, 2008).

69 An EAF consuming only 340 kWh/t of steel has power of 125–130 MW, and its daily operation (40 heats of 120 t) will need 1.63 GWh of electricity. Using the average annual US household electricity consumption of about 29 kWh/day and average household size of 2.52 people, this works out to an equivalent of about 56,000 households or 141,000 people.

70 "Alang, Gujarat: The World's Biggest Ship Breaking Yard & A Dangerous Environmental Time Bomb," Marine Insight (March 2019), https://www.marineinsight.com/environment/alang-gujarat-the-world's-biggest-ship-breaking-yard-a-dangerous-environmental-time-bomb/. In March 2020, Google satellite view showed more than 70 vessels and drilling rigs in various stages of dismantling on the Alang beaches between P. Rajesh Shipbreaking at the southern end and Rajendra Shipbreakers, about 10 km to the northwest.

71 Concrete Reinforcing Steel Institute, "Recycled materials" (2020), https://www.crsi.org/index.cfm/architecture/recycling.

72 Bureau of International Recycling, *World Steel Recycling in Figures 2014–2018* (Brussels: Bureau of International Recycling, 2019).

73 World Steel Association, *Steel in Figures 2019* (Brussels: World Steel Association, 2019).

74 For the long history of blast furnaces, see Smil, *Still the Iron Age*. For the construction and operation of modern blast furnaces, see

M. Geerdes et al., *Modern Blast Furnace Ironmaking* (Amsterdam: IOS Press, 2009); I. Cameron et al., *Blast Furnace Ironmaking* (Amsterdam: Elsevier, 2019).

75 Invention and diffusion of basic oxygen furnaces is traced in W. Adams and J. B. Dirlam, "Big steel, invention, and innovation," *Quarterly Journal of Economics* 80 (1966), pp. 167–189; T. W. Miller et al., "Oxygen steelmaking processes," in D. A. Wakelin, ed., *The Making, Shaping and Treating of Steel: Ironmaking Volume* (Pittsburgh, PA: The AISE Foundation, 1998), pp. 475–524; J. Stubbles, "EAF steelmaking—past, present and future," *Direct from MIDREX* 3 (2000), pp. 3–4.

76 World Steel Association, "Energy use in the steel industry" (2019), https://www.worldsteel.org/en/dam/jcr:f07b864c-908e-4229-9f92-669f1c3abf4c/fact_energy_2019.pdf.

77 For historical trends, see Smil, *Still the Iron Age*; US Energy Information Administration, "Changes in steel production reduce energy intensity" (2016), https://www.eia.gov/todayinenergy/detail.php?id=27292.

78 World Steel Association, *Steel's Contribution to a Low Carbon Future and Climate Resilient Societies* (Brussels: World Steel Association, 2020); H. He et al., "Assessment on the energy flow and carbon emissions of integrated steelmaking plants," *Energy Reports* 3 (2017), pp. 29–36.

79 J. P. Saxena, *The Rotary Cement Kiln: Total Productive Maintenance, Techniques and Management* (Boca Raton, FL: CRC Press, 2009).

80 V. Smil, "Concrete facts," *Spectrum IEEE* (March 2020), pp. 20–21; National Concrete Ready Mix Associations, *Concrete CO_2 Fact Sheet* (2008).

81 F.-J. Ulm, "Innovationspotenzial Beton: Von Atomen zur Grünen Infrastruktur," *Beton- und Stahlbetonbauer* 107 (2012), pp. 504–509.

82 Modern wooden buildings have been getting taller, but they do not use plain lumber but rather much stronger cross-laminated timber (CLT), a proprietary engineered material prefabricated from several (3, 5, 7, or 9) layers of kiln-dried lumber that are laid flat and glued together: https://cwc.ca/how-to-build-with-wood/wood-products/mass-timber/cross-laminated-timber-clt/. In 2020,

the world's tallest (85.4 m) CLT building was Mjøstårnet by Voll Arkitekter in Brumunddal, Norway, a multipurpose (apartments, a hotel, offices, a restaurant, a swimming pool) structure completed in 2019: https://www.dezeen.com/2019/03/19/mjostarne-worlds-tallest-timber-tower-voll-arkitekter-norway/.

83 F. Lucchini, *Pantheon—Monumenti dell' Architettura* (Roma: Nuova Italia Scientifica, 1966).

84 A. J. Francis, *The Cement Industry, 1796–1914: A History* (Newton Abbot: David and Charles, 1978).

85 Smil, "Concrete facts."

86 J.-L. Bosc, *Joseph Monier et la naissance du ciment armé* (Paris: Editions du Linteau, 2001); F. Newby, ed., *Early Reinforced Concrete* (Burlington, VT: Ashgate, 2001).

87 American Society of Civil Engineers, "Ingalls building" (2020), https://www.asce.org/project/ingalls-building/; M. M. Ali, "Evolution of Concrete Skyscrapers: from Ingalls to Jin Mao," *Electronic Journal of Structural Engineering* 1 (2001), pp. 2–14.

88 M. Peterson, "Thomas Edison's Concrete Houses," *Invention & Technology* 11/3 (1996), pp. 50–56.

89 D. P. Billington, *Robert Maillart and the Art of Reinforced Concrete* (Cambridge, MA: MIT Press, 1990).

90 B. B. Pfeiffer and D. Larkin, *Frank Lloyd Wright: The Masterworks* (New York: Rizzoli, 1993).

91 E. Freyssinet, *Un amour sans limite* (Paris: Editions du Linteau, 1993).

92 *Sydney Opera House: Utzon Design Principles* (Sydney: Sydney Opera House, 2002).

93 History of Bridges, "The World's Longest Bridge—Danyang–Kunshan Grand Bridge" (2020), http://www.historyofbridges.com/famous-bridges/longest-bridge-in-the-world/.

94 US Geological Survey, "Materials in Use in U.S. Interstate Highways" (2006), https://pubs.usgs.gov/fs/2006/3127/2006-3127.pdf.

95 Associated Engineering, "New runway and tunnel open skies and roads at Calgary International Airport" (June 2015).

96 Among many books about the Hoover Dam, the eyewitness accounts contained in the following stand out: A. J. Dunar and

D. McBride, *Building Hoover Dam: An Oral History of the Great Depression* (Las Vegas: University of Nevada Press, 2016).
97 Power Technology, "Three Gorges Dam Hydro Electric Power Plant, China" (2020), https://www.power-technology.com/projects/gorges/.
98 Data on the production, trade, and consumption of American cement are available from the annual summaries published by the US Geological Survey. The 2020 edition: US Geological Survey, *Mineral Commodity Summaries 2020*, https://pubs.usgs.gov/periodicals/mcs2020/mcs2020.pdf.
99 At 320 million tons, India's 2019 production, the world's second largest, is only 15 percent of the Chinese total: USGS, "Cement" (2020), https://pubs.usgs.gov/periodicals/mcs2020/mcs2020-cement.pdf.
100 N. Delatte, ed., *Failure, Distress and Repair of Concrete Structures* (Cambridge: Woodhead Publishing, 2009).
101 D. R. Wilburn and T. Goonan, *Aggregates from Natural and Recycled Sources* (Washington, DC: USGS, 2013).
102 American Society of Civil Engineers, *2017 Infrastructure Report Card*, https://www.infrastructurereportcard.org/.
103 C. Kenny, "Paving Paradise," *Foreign Policy* (Jan/Feb 2012), pp. 31–32.
104 Abandoned concrete structures around the world now include almost any kind of building, from nuclear submarine bases to nuclear reactors (each one of these can be found in Ukraine), and from railway stations and large sports stadiums to theaters and monuments.
105 Calculated from official Chinese data published annually in the country's *China Statistical Yearbook*. The latest edition is available at: http://www.stats.gov.cn/tjsj/ndsj/2019/indexeh.htm.
106 M. P. Mills, *Mines, Minerals, and "Green" Energy: A Reality Check* (New York: Manhattan Institute, 2020).
107 V. Smil, "What I see when I see a wind turbine," *IEEE Spectrum* (March 2016), p. 27.
108 H. Berg and M. Zackrisson, "Perspectives on environmental and cost assessment of lithium metal negative electrodes in electric vehicle traction batteries," *Journal of Power Sources* 415 (2019), pp.

83–90; M. Azevedo et al., *Lithium and Cobalt: A Tale of Two Commodities* (New York: McKinsey & Company, 2018).

109 C. Xu et al., "Future material demand for automotive lithium-based batteries," *Communications Materials* 1 (2020), p. 99.

4. Understanding Globalization: Engines, Microchips, and Beyond

1 For the origins of iPhone parts, see "Here's where all the components of your iPhone come from," Business Insider, https://i.insider.com/570d5092dd089568298b4978; and see the actual parts at: "iPhone 11 Pro Max Teardown," iFixit (September 2019), https://www.ifixit.com/Teardown/iPhone+11+Pro+Max+Teardown/126000.

2 Nearly 1.1 million foreign students were enrolled at US universities and colleges during the academic year 2018/2019, making up 5.5 percent of the total and contributing $44.7 billion to the US economy: Open Doors 2019 Data Release, https://opendoorsdata.org/annual-release/.

3 Nothing conveys the pre-COVID plague of overtourism as do the images of major tourist destinations covered by masses of people: just search for "overtourism" and click on Images.

4 World Trade Organization, *Highlights of World Trade* (2019), https://www.wto.org/english/res_e/statis_e/wts2019_e/wts2019chapter02_e.pdf.

5 World Bank, "Foreign direct investment, net inflows" (accessed 2020), https://data.worldbank.org/indicator/BX.KLT.DINV.CD.WD; A. Debnath and S. Barton, "Global currency trading surges to $6.6 trillion-a-day market," GARP (September 2019), https://www.garp.org/#!/risk-intelligence/all/all/a1Z1W000003mKKPUA2.

6 V. Smil, "Data world: Racing toward yotta," *IEEE Spectrum* (July 2019), p. 20. For details on the unit sub-multiples, see the Appendix.

7 Peterson Institute for International Economics, "What is globalization?" (accessed 2020), https://www.piie.com/microsites/globalization/what-is-globalization.

8 W. J. Clinton, *Public Papers of the Presidents of the United States: William J. Clinton, 2000–2001* (Best Books, 2000).

9 World Bank, "Foreign direct investment, net inflows."

10 Obviously, lack of personal freedoms or high levels of corruption are no obstacles to large investment inflows. China's freedom score is 10 and India's is 71 out of a possible 100 (Canada's is 98), and China shares the high ranking on the corruption perception index (80, compared to Finland's 3) with India: Freedom House, "Countries and territories" (accessed 2020), https://freedomhouse.org/countries/freedom-world/scores; Transparency International, "Corruption perception index" (accessed 2020), https://www.transparency.org/en/cpi/2020/index/nzl.

11 G. Wu, "Ending poverty in China: What explains great poverty reduction and a simultaneous increase in inequality in rural areas?" World Bank Blogs (October 2016), https://blogs.worldbank.org/eastasiapacific/ending-poverty-in-china-what-explains-great-poverty-reduction-and-a-simultaneous-increase-in-inequality-in-rural-areas.

12 Here is just a small selection of some noteworthy contributions: J. E. Stieglitz, *Globalization and Its Discontents* (New York: W.W. Norton, 2003); G. Buckman, *Globalization: Tame It or Scrap It?: Mapping the Alternatives of the Anti-Globalization Movement* (London: Zed Books, 2004); M. Wolf, *Why Globalization Works* (New Haven, CT: Yale University Press, 2005); P. Marber, "Globalization and its contents," *World Policy Journal* 21 (2004), pp. 29–37; J. Bhagvati, *In Defense of Globalization* (Oxford: Oxford University Press, 2007); J. Miśkiewicz and M. Ausloos, "Has the world economy reached its globalization limit?" *Physica A: Statistical Mechanics and its Applications* 389 (2009), pp. 797–806; L. J. Brahm, *The Anti-Globalization Breakfast Club: Manifesto for a Peaceful Revolution* (Chichester: John Wiley, 2009); D. Rodrik, *The Globalization Paradox: Democracy and the Future of the World Economy* (New York: W.W. Norton, 2011); R. Baldwin, *The Great Convergence: Information Technology and the New Globalization* (Cambridge, MA: Belknap Press, 2016).

13 J. Yellin et al., "New evidence on prehistoric trade routes: The obsidian evidence from Gilat, Israel," *Journal of Field Archaeology* 23 (2013), pp. 361–368.

14 Cassius Dio, *Romaika* LXVIII:29: "Then he came to the ocean itself, and when he had learned its nature and had seen a ship sailing to India, he said: 'I should certainly have crossed over to the Indi, too, if I were still young.' For he began to think about the Indi and was curious about their affairs, and he counted Alexander a lucky man." (E. Cary translation).

15 V. Smil, *Why America is Not a New Rome* (Cambridge, MA: MIT Press, 2008).

16 J. Keay, *The Honourable Company: A History of the English East India Company* (London: Macmillan, 1994); F. S. Gaastra, *The Dutch East India Company* (Zutpen: Walburg Press, 2007).

17 Porters with heavy loads (50–70 kg) in mountainous terrain could not do more than 9–11 km a day; with lighter loads (35–40 kg) they could cover up to 24 km a day, the same distance as horse caravans: N. Kim, *Mountain Rivers, Mountain Roads: Transport in Southwest China, 1700–1850* (Leiden: Brill, 2020), p. 559.

18 J. R. Bruijn et al., *Dutch-Asiatic Shipping in the 17th and 18th Centuries* (The Hague: Martinus Nijhoff, 1987).

19 J. Lucassen, "A multinational and its labor force: The Dutch East India Company, 1595–1795," *International Labor and Working-Class History* 66 (2004), pp. 12–39.

20 C. Mukerji, *From Graven Images: Patterns of Modern Materialism* (New York: Columbia University Press, 1983).

21 W. Franits, *Dutch Seventeenth-Century Genre Painting* (New Haven, CT: Yale University Press, 2004); D. Shawe-Taylor and Q. Buvelot, *Masters of the Everyday: Dutch Artists in the Age of Vermeer* (London: Royal Collection Trust, 2015).

22 W. Fock, "Semblance or Reality? The Domestic Interior in Seventeenth-Century Dutch Genre Painting," in M. Westermann, ed., *Art & Home: Dutch Interiors in the Age of Rembrandt* (Zwolle: Waanders, 2001), pp. 83–101.

23 J. de Vries, "Luxury in the Dutch Golden Age in theory and practice," in M. Berg and E. Eger, eds., *Luxury in the Eighteenth Century* (London: Palgrave Macmillan, 2003), pp. 41–56.

24 D. Hondius, "Black Africans in seventeenth century Amsterdam," *Renaissance and Reformation* 31 (2008), pp. 87–105; T. Moritake,

"Netherlands and tea," World Green Tea Association (2020), http://www.o-cha.net/english/teacha/history/netherlands.html.

25 A. Maddison, "Dutch income in and from Indonesia 1700–1938," *Modern Asia Studies* 23 (1989), pp. 645–670.

26 R. T. Gould, *Marine Chronometer: Its History and Developments* (New York: ACC Art Books, 2013).

27 C. K. Harley, "British shipbuilding and merchant shipping: 1850–1890," *Journal of Economic History* 30/1 (1970), pp. 262–266.

28 R. Knauerhase, "The compound steam engine and productivity: Changes in the German merchant marine fleet, 1871–1887," *Journal of Economic History* 28/3 (1958), pp. 390–403.

29 C. L. Harley, "Steers afloat: The North Atlantic meat trade, liner predominance, and freight rates, 1870–1913," *Journal of Economic History* 68/4 (2008), pp. 1028–1058.

30 For the history of telegraph, see F. B. Jewett, *100 Years of Electrical Communication in the United States* (New York: American Telephone and Telegraph, 1944); D. Hochfelder, *The Telegraph in America, 1832–1920* (Baltimore, MD: Johns Hopkins University Press, 2013); R. Wenzlhuemer, *Connecting the Nineteenth-Century World. The Telegraph and Globalization* (Cambridge: Cambridge University Press, 2012).

31 For the early history of the telephone, see H. N. Casson, *The History of the Telephone* (Chicago: A. C. McClurg & Company, 1910); E. Garcke, "Telephone," in *Encyclopaedia Britannica*, 11th edn, vol. 26 (Cambridge: Cambridge University Press, 1911), pp. 547–557.

32 Smil, *Creating the Twentieth Century*.

33 G. Federico and A. Tena-Junguito, "World trade, 1800–1938: a new synthesis," *Revista de Historia Económica / Journal of Iberian and Latin America Economic History* 37/1 (2019); CEPII, "Databases," http://www.cepii.fr/CEPII/en/bdd_modele/bdd.asp; M. J. Klasing and P. Milionis, "Quantifying the evolution of world trade, 1870–1949," *Journal of International Economics* 92/1 (2014), pp. 185–197. For a history of "steam globalization," see J. Darwin, *Unlocking the World: Port Cities and Globalization in the Age of Steam, 1830–1930* (London: Allen Lane, 2020).

34 US Department of Homeland Security, "Total immigrants by decade," http://teacher.scholastic.com/activities/immigration/pdfs/by_decade/decade_line_chart.pdf.

35 The rise of 19th-century tourism is described in P. Smith, *The History of Tourism: Thomas Cook and the Origins of Leisure Travel* (London: Psychology Press, 1998); E. Zuelow, *A History of Modern Tourism* (London: Red Globe Press, 2015).

36 Lenin lived and traveled in Western Europe (France, Switzerland, England, Germany, and Belgium) and Austrian Poland between July 1900 and November 1905, and then between December 1907 and April 1917: R. Service, *Lenin: A Biography* (Cambridge, MA: Belknap Press, 2002).

37 Smil, *Prime Movers of Globalization*.

38 F. Oppel, ed., *Early Flight* (Secaucus, NJ: Castle, 1987); B. Gunston, *Aviation: The First 100 Years* (Hauppauge, NY: Barron's, 2002).

39 M. Raboy, *Marconi: The Man Who Networked the World* (Oxford: Oxford University Press, 2018); H. G. J. Aitkin, *The Continuous Wave: Technology and the American Radio, 1900–1932* (Princeton, NJ: Princeton University Press, 1985).

40 Smil, *Prime Movers of Globalization*.

41 J. J. Bogert, "The new oil engines," *The New York Times* (September 26, 1912), p. 4.

42 E. Davies et al., *Douglas DC-3: 60 Years and Counting* (Elk Grove, CA: Aero Vintage Books, 1995); M. D. Klaás, *Last of the Flying Clippers* (Atglen, PA: Schiffer Publishing, 1998); "Pan Am across the Pacific," Pan Am Clipper Flying Boats (2009), https://www.clipperflyingboats.com/transpacific-airline-service.

43 M. Novak, "What international air travel was like in the 1930s," Gizmodo (2013), https://paleofuture.gizmodo.com/what-international-air-travel-was-like-in-the-1930s-1471258414.

44 J. Newman, "Titanic: Wireless distress messages sent and received April 14–15, 1912," Great Ships (2012), https://greatships.net/distress.

45 A. K. Johnston et al., *Time and Navigation* (Washington, DC: Smithsonian Books, 2015).

46 For a graph of adoption rates of new devices, see D. Thompson, "The 100-year march of technology in 1 graph," *The Atlantic* (April 2012), https://www.theatlantic.com/technology/archive/2012/04/the-100-year-march-of-technology-in-1-graph/255573/.
47 V. Smil, *Made in the USA: The Rise and Retreat of American Manufacturing* (Cambridge, MA: MIT Press, 2013).
48 S. Okita, "Japan's Economy and the Korean War," *Far Eastern Survey* 20 (1951), pp. 141–144.
49 Historical statistics (national and global) of steel, cement, and ammonia (nitrogen) production are available at: US Geological Survey, "Commodity statistics and information," https://www.usgs.gov/centers/nmic/commodity-statistics-and-information. For plastic production, see R. Geyer et al., "Production, use, and fate of all plastics ever made," *Science Advances* 3/7 (2017), e1700782.
50 R. Solly, *Tanker: The History and Development of Crude Oil Tankers* (Barnsley: Chatham Publishing, 2007).
51 United Nations, *World Energy Supplies in Selected Years 1929–1950* (New York: UN, 1952); British Petroleum, *Statistical Review of World Energy*.
52 P. G. Noble, "A short history of LNG shipping, 1959–2009," SNAME (2009).
53 M. Levinson, *The Box* (Princeton, NJ: Princeton University Press, 2006); Smil, *Prime Movers of Globalization*.
54 For the rise of imports and the declining share of Detroit cars in the US car market, see Smil, *Made in the USA*.
55 Germany's MAN (Maschinenfabrik-Augsburg-Nürnberg) led the technical advances in post-Second World War diesel engines, but now the largest machines are designed by Finnish Wärtsilä and made in Asia (Japan, South Korea, China): https://www.wartsila.com/marine/build/engines-and-generating-sets/diesel-engines (accessed 2020).
56 Smil, *Prime Movers of Globalization*, pp. 79–108.
57 G. M. Simons, *Comet! The World's First Jet Airliner* (Philadelphia: Casemate, 2019).
58 E. E. Bauer, *Boeing: The First Century* (Enumclaw, WA: TABA Publishers, 2000); A. Pelletier, *Boeing: The Complete Story* (Sparkford: Haynes Publishing, 2010).

59 More books have been published about the 747 than about any other commercial aircraft in history. J. Sutter and J. Spenser, *747: Creating the World's First Jumbo Jet and Other Adventures from a Life in Aviation* (Washington, DC: Smithsonian, 2006). For an inside look, see C. Wood, *Boeing 747 Owners' Workshop Manual* (London: Zenith Press, 2012).

60 "JT9D Engine," Pratt & Whitney (accessed 2020), https://prattwhitney.com/products-and-services/products/commercial-engines/jt9d. For details on turbofans, see N. Cumpsty, *Jet Propulsion* (Cambridge: Cambridge University Press, 2003); A. Linke-Diesinger, *Systems of Commercial Turbofan Engines* (Berlin: Springer, 2008).

61 E. Lacitis, "50 years ago, the first 747 took off and changed aviation," *The Seattle Times* (February 2019).

62 S. McCartney, *ENIAC* (New York: Walker & Company, 1999).

63 T. R. Reid, *The Chip* (New York: Random House, 2001); C. Lécuyer and D. C. Brock, *Makers of the Microchip* (Cambridge: MIT Press, 2010).

64 "The story of the Intel 4044," Intel (accessed 2020), https://www.intel.com/content/www/us/en/history/museum-story-of-intel-4004.html.

65 World Bank, "Export of goods and services (percentage of GDP)" (accessed 2020), https://data.worldbank.org/indicator/ne.exp.gnfs.zs.

66 United Nations, *World Economic Survey, 1975* (New York: UN, 1976).

67 S. A. Camarota, *Immigrants in the United States, 2000* (Center for Immigration Studies, 2001), https://cis.org/Report/Immigrants-United-States-2000.

68 P. Nolan, *China and the Global Business Revolution* (London: Palgrave, 2001); L. Brandt et al., eds., *China's Great Transformation* (Cambridge: Cambridge University Press, 2008).

69 S. Kotkin, *Armageddon Averted: The Soviet Collapse, 1970–2000* (Oxford: Oxford University Press, 2008).

70 C. VanGrasstek, *The History and Future of the World Trade Organization* (Geneva: WTO, 2013).

71 World Bank, "GDP per capita growth (annual percent)—India" (accessed 2020), https://data.worldbank.org/indicator/NY.GDP.PCAP.KD.ZG?locations=IN.

72 World Trade Organization, *World Trade Statistical Review 2019* (Geneva: WTO, 2019), https://www.wto.org/english/res_e/statis_e/wts2019_e/wts2019_e.pdf.

73 World Bank, "Trade share (percent of GDP)" (accessed 2020), https://data.worldbank.org/indicator/ne.trd.gnfs.zs.

74 World Bank, "Foreign direct investment, net outflows (percent of GDP)" (accessed 2020), https://data.worldbank.org/indicator/BM.KLT.DINV.WD.GD.ZS.

75 S. Shulgin et al., "Measuring globalization: Network approach to countries' global connectivity rates and their evolution in time," *Social Evolution & History* 18/1 (2019), pp. 127–138.

76 United Nations Conference on Trade and Development, *Review of Maritime Transport, 1975* (New York: UNCTAD, 1977); *Review of Maritime Transport, 2019* (New York: UNCTAD, 2020); *50 Years of Review of Maritime Transport, 1968–2018* (New York: UNCTAD, 2018).

77 Maersk, "About our group," https://web.archive.org/web/20071012231026/http://about.maersk.com/en; Mediterranean Shipping Company, "Gülsün Class Ships" (accessed 2020), https://www.msc.com/tha/about-us/new-ships.

78 International Air Transport Association, *World Air Transport Statistics* (Montreal: IATA, 2019), and the past volumes of this annual publication.

79 World Tourism Organization, "Tourism statistics" (accessed 2020), https://www.e-unwto.org/toc/unwtotfb/current.

80 K. Koens et al., *Overtourism? Understanding and Managing Urban Tourism Growth beyond Perceptions* (Madrid: World Tourism Organization, 2018).

81 G. E. Moore, "Cramming more components onto integrated circuits," *Electronics* 38/8 (1965), pp. 114–117; "Progress in digital integrated electronics," *Technical Digest, IEEE International Electron Devices Meeting* (1975), pp. 11–13; "No exponential is forever: but 'Forever' can be delayed!", paper presented at Solid-State

Circuits Conference, San Francisco (2003); Intel, "Moore's law and Intel innovation" (accessed 2020), http://www.intel.com/content/www/us/en/history/museum-gordon-moore-law.html.

82 C. Tung et al., *ULSI Semiconductor Technology Atlas* (Hoboken, NJ: Wiley-Interscience, 2003).

83 J. V. der Spiegel, "ENIAC-on-a-chip," Moore School of Electrical Engineering (1995), https://www.seas.upenn.edu/~jan/eniacproj.html.

84 H. Mujtaba, "AMD 2nd gen EPYC Rome processors feature a gargantuan 39.54 billion transistors, IO die pictured in detail," WCCFTech (October 2019), https://wccftech.com/amd-2nd-gen-epyc-rome-iod-ccd-chipshots-39-billion-transistors/.

85 P. E. Ceruzzi, *GPS* (Cambridge, MA: MIT Press, 2018); A. K. Johnston et al., *Time and Navigation* (Washington, DC: Smithsonian Books, 2015).

86 MarineTraffic, https://www.marinetraffic.com.

87 Flightradar24, https://www.flightradar24.com; Flight Aware, https://flightaware.com/live/.

88 For example, the normal flight path (following the great circle route) from Frankfurt (FRA) to Chicago (ORD) passes south of Greenland's southernmost tip (see: Great Circle Mapper, http://www.gcmap.com/mapui?P=FRA-ORD). But when a strong jet stream moves in, the trajectory shifts northward and planes overfly the island's glaciers.

89 The most notable recent disruption of flights was due to the eruption of Iceland's Eyjafjallajökull in April and May 2010: BGS Research, "Eyjafjallajökull eruption, Iceland," British Geological Survey (accessed 2020), https://www.bgs.ac.uk/research/volcanoes/icelandic_ash.html.

90 M. J. Klasing and P. Milionis, "Quantifying the evolution of world trade, 1870–1949," *Journal of International Economics* 92 (2014), pp. 185–197.

91 For a map of food self-sufficiency ratio, see Food and Agriculture Organization, "Food self-sufficiency and international trade: a false dichotomy?" in *The State of Agricultural Markets IN DEPTH 2015–16* (Rome: FAO, 2016), http://www.fao.org/3/a-i5222e.pdf.

92 The Internet helpfully suggests 10, 13, 20, 23, 50 or 100 bucket list destinations—just search for "Bucket list places to visit."

93 The US and EU's declining shares in global manufacturing are reviewed in M. Levinson, *U.S. Manufacturing in International Perspective* (Congressional Research Service, 2018), https://fas.org/sgp/crs/misc/R42135.pdf; and R. Marschinski and D. Martínez-Turégano, "The EU's shrinking share in global manufacturing: a value chain decomposition analysis," *National Institute Economic Review* 252 (2020), R19–R32.

94 Despite having a chronic large trade deficit with China, in 2019 Canada's imports included nearly half a billion dollars' worth of paper, paperboard, and pulp—and yet Canada's per capita area of naturally regenerating forests is about 90 times larger than in China: FAO, *Global Forest Resources Assessment 2020*, http://www.fao.org/3/ca9825en/CA9825EN.pdf.

95 A. Case and A. Deaton, *Deaths of Despair and the Future of Capitalism* (Princeton, NJ: Princeton University Press, 2020).

96 S. Lund et al., *Globalization in Transition: The Future of Trade and Value Chains* (Washington, DC: McKinsey Global Institute, 2019).

97 OECD, *Trade Policy Implications of Global Value Chains* (Paris: OECD, 2020).

98 A. Ashby, "From global to local: reshoring for sustainability," *Operations Management Research* 9/3–4 (2016), pp. 75–88; O. Butzbach et al., "Manufacturing discontent: National institutions, multinational firm strategies, and anti-globalization backlash in advanced economies," *Global Strategy Journal* 10 (2019), pp. 67–93.

99 OECD, "COVID-19 and global value chains: Policy options to build more resilient production networks" (June 2020); UNCTAD, *World Investment Report 2020* (New York: UNCTAD, 2020); Swiss Re Institute, "De-risking global supply chains: Rebalancing to strengthen resilience," *Sigma* 6 (2020); A. Fish and H. Spillane, "Reshoring advanced manufacturing supply chains to generate good jobs," Brookings (July 2020), https://www.brookings.edu/research/reshoring-advanced-manufacturing-supply-chains-to-generate-good-jobs/.

100 V. Smil, "History and risk," *Inference* 5/1 (April 2020). Six months into the COVID pandemic, dire shortages of PPE still persisted in American hospitals: D. Cohen, "Why a PPE shortage still plagues America and what we need to do about it," CNBC (August 2020), https://www.cnbc.com/2020/08/22/coronavirus-why-a-ppe-shortage-still-plagues-the-us.html.

101 P. Haddad, "Growing Chinese transformer exports cause concern in U.S.," Power Transformer News (May 2019), https://www.powertransformernews.com/2019/05/02/growing-chinese-transformer-exports-cause-concern-in-u-s/.

102 N. Stonnington, "Why reshoring U.S. manufacturing could be the wave of the future," *Forbes* (September 9, 2020); M. Leonard, "64 percent of manufacturers say reshoring is likely following pandemic: survey," Supply Chain Dive (May 2020), https://www.supplychaindive.com/news/manufacturing-reshoring-pandemic-thomas/577971/.

5. *Understanding Risks: From Viruses to Diets to Solar Flares*

1 A. de Waal, "The end of famine? Prospects for the elimination of mass starvation by political action," *Political Geography* 62 (2017), pp. 184–195.

2 On the impact of more frequent handwashing, see Global Handwashing Partnership, "About handwashing" (accessed 2020), https://globalhandwashing.org/about-handwashing/. The risks of CO poisoning used to be particularly high in cold climates where woodstoves were the only source of heat: J. Howell et al., "Carbon monoxide hazards in rural Alaskan homes," *Alaska Medicine* 39 (1997), pp. 8–11. With so many kinds of inexpensive CO detectors (the first ones were introduced commercially during the early 1990s), there is no excuse for any fatalities from incomplete in-house combustion.

3 There is probably no other design of a comparative simplicity than the three-point automotive seat belt (first by Nils Ivar Bohlin for Volvo in 1959) that could be credited with saving so many lives

and preventing many more grievous injuries—and doing so at such a low cost. Justifiably, in 1985 the German Patent Office ranked it among the eight most important innovations of the preceding 100 years. N. Bohlin, "A statistical analysis of 28,000 accident cases with emphasis on occupant restraint value," SAE Technical Paper 670925 (1967); T. Borroz, "Strapping success: The 3-point seatbelt turns 50," *Wired* (August 2009).

4 This matter has been a long-running irritant for Japan's external relations. The country had repeatedly refused to sign The Hague Convention on the Civil Aspects of International Child Abduction (signed in 1980, in force since December 1, 1983): Convention on the Civil Aspects of International Child Abduction, https://assets.hcch.net/docs/e86d9f72-dc8d-46f3-b3bf-e102911c8532.pdf. And although it did finally sign it in 2014, few American or European partners have succeeded in reclaiming their parental rights.

5 On the decline of violent conflicts, see J. R. Oneal, "From realism to the liberal peace: Twenty years of research on the causes of war," in G. Lundestad, ed., *International Relations Since the End of the Cold War: Some Key Dimensions* (Oxford: Oxford University Press, 2012), pp. 42–62; S. Pinker, "The decline of war and conceptions of human nature," *International Studies Review* 15/3 (2013), pp. 400–405.

6 National Cancer Institute, "Asbestos exposure and cancer risk" (accessed 2020), https://www.cancer.gov/about-cancer/; American Cancer Society, "Talcum powder and cancer" (accessed 2020), https://www.cancer.org/cancer/cancer-causes/talcum-powder-and-cancer.html; J. Entine, *Scared to Death: How Chemophobia Threatens Public Health* (Washington, DC: American Council on Science and Health, 2011). On global warming there is a wide choice of recent apocalyptic books, and the challenge will be considered in the next two chapters.

7 S. Knobler et al., *Learning from SARS: Preparing for the Next Disease Outbreak—Workshop Summary* (Washington, DC: National Academies Press, 2004); D. Quammen, *Ebola: The Natural and Human History of a Deadly Virus* (New York: W. W. Norton, 2014).

8 Risk literature is now enormous, with many specialized branches: books and papers on business risk management are particularly numerous, followed by publications on natural hazards. The three leading periodicals are *Risk Analysis*, *Journal of Risk Research*, and *Journal of Risk*.

9 For the story of human evolution during the Paleolithic era, see F. J. Ayala and C. J. Cela-Cond, *Processes in Human Evolution: The Journey from Early Hominins to Neandertals and Modern Humans* (New York: Oxford University Press, 2017). For the claims of the "Paleolithic" diet's efficacy, see https://thepaleodiet.com/. For an impartial review of the diet, consult: Harvard T. H. Chan School of Public Health, "Diet review: paleo diet for weight loss" (accessed 2020), https://www.hsph.harvard.edu/nutritionsource/healthy-weight/diet-reviews/paleo-diet/. There is no shortage of books that promise not only to turn you vegetarian, or even vegan, but, "quite literally, to save the world." For just two much-publicized takes, see J. M. Masson, *The Face on Your Plate: The Truth About Food* (New York: W.W. Norton, 2010); and J. S. Foer, *We Are the Weather: Saving the Planet Begins at Breakfast* (New York: Farrar, Straus and Giroux, 2019).

10 E. Archer et al., "The failure to measure dietary intake engendered a fictional discourse on diet-disease relations," *Frontiers in Nutrition* 5 (2019), p. 105. For the most extensive, and also the most accusatory, exchange of opinions regarding modern prospective dietary studies, see the four sets of comments starting with E. Archer et al., "Controversy and debate: Memory-Based Methods Paper 1: The fatal flaws of food frequency questionnaires and other memory-based dietary assessment methods," *Journal of Clinical Epidemiology* 104 (2018), pp. 113–124.

11 The most far-reaching controversy has concerned the role of dietary fats and cholesterol in heart disease. For the original claims, see American Heart Association, "Dietary guidelines for healthy American adults," *Circulation* 94 (1966), pp. 1795–1800; A. Keys, *Seven Countries: A Multivariate Analysis of Death and Coronary Heart Disease* (Cambridge, MA: Harvard University Press, 1980). For their critique and reversals of earlier claims, see A. F. La Berge,

"How the ideology of low fat conquered America," *Journal of the History of Medicine and Allied Sciences* 63/2 (2008), pp. 139–177; R. Chowdhury et al., "Association of dietary, circulating, and supplement fatty acids with coronary risk: a systematic review and meta-analysis," *Annals of Internal Medicine* 160/6 (2014), pp. 398–406; R. J. De Souza et al., "Intake of saturated and trans unsaturated fatty acids and risk of all cause mortality, cardiovascular disease, and type 2 diabetes: systematic review and meta-analysis of observational studies," *British Medical Journal* (2015); M. Dehghan et al., "Associations of fats and carbohydrate intake with cardiovascular disease and mortality in 18 countries from five continents (PURE): a prospective cohort study," *The Lancet* 390/10107 (2017), pp. 2050–2062; American Heart Association, "Dietary cholesterol and cardiovascular risk: A science advisory from the American Heart Association," *Circulation* 141 (2020), e39–e53.

12 Life expectancies as a five-year average between 1950 and 2020 are available for all countries and regions at: United Nations, *World Population Prospects 2019*, https://population.un.org/wpp/Download/Standard/Population/.

13 Detailed Japanese historical statistics document this trend. Statistics Bureau, Japan, *Historical Statistics of Japan* (Tokyo: Statistics Bureau, 1996).

14 H. Toshima et al., eds., *Lessons for Science from the Seven Countries Study: A 35-Year Collaborative Experience in Cardiovascular Disease Epidemiology* (Berlin: Springer, 1994).

15 For more on the consumption of total and added sugar in the US and Japan, see S. A. Bowman et al., *Added Sugars Intake of Americans: What We Eat in America, NHANES 2013–2014* (May 2017); A. Fujiwara et al., "Estimation of starch and sugar intake in a Japanese population based on a newly developed food composition database," *Nutrients* 10 (2018), p. 1474.

16 Good introductions include M. Ashkenazi and J. Jacob, *The Essence of Japanese Cuisine* (Philadelphia: University of Philadelphia Press, 2000); K. J. Cwiertka, *Modern Japanese Cuisine* (London: Reaktion Books, 2006); E. C. Rath and S. Assmann, eds., *Japanese Foodways: Past & Present* (Urbana, IL: University of Illinois Press, 2010).

17 Apparent consumption rates in Spain are from: Fundación Foessa, *Estudios sociológicos sobre la situación social de España, 1975* (Madrid: Editorial Euramerica, 1976), p. 513; Ministerio de Agricultura, Pesca y Alimentación, *Informe del Consume Alimentario en España 2018* (Madrid: Ministerio de Agricultura, Pesca y Alimentación, 2019).

18 Comparisons based on: FAO, "Food Balances" (accessed 2020), http://www.fao.org/faostat/en/#data/FBS.

19 For CVD mortality, see L. Serramajem et al., "How could changes in diet explain changes in coronary heart disease mortality in Spain—The Spanish Paradox," *American Journal of Clinical Nutrition* 61 (1995), S1351–S1359; OECD, *Cardiovascular Disease and Diabetes: Policies for Better Health and Quality of Care* (June 2015). For life expectancy, see United Nations, *World Population Prospects 2019*.

20 C. Starr, "Social benefit versus technological risk," *Science* 165 (1969), pp. 1232–1238.

21 According to a detailed quantitative risk assessment, tobacco smoke contains 18 harmful and potentially harmful constituents: K. M. Marano et al., "Quantitative risk assessment of tobacco products: A potentially useful component of substantial equivalence evaluations," *Regulatory Toxicology and Pharmacology* 95 (2018), pp. 371–384.

22 M. Davidson, "Vaccination as a cause of autism—myths and controversies," *Dialogues in Clinical Neuroscience* 19/4 (2017), pp. 404–407; J. Goodman and F. Carmichael, "Coronavirus: Bill Gates 'microchip' conspiracy theory and other vaccine claims fact-checked," BBC News (May 29, 2020).

23 In early September 2020, two-thirds of Americans said they will not take the COVID-19 vaccine when it becomes available: S. Elbeshbishi and L. King, "Exclusive: Two-thirds of Americans say they won't get COVID-19 vaccine when it's first available, USA TODAY/Suffolk Poll shows," USA Today (September 2020).

24 Comprehensive reports on the health consequences of the two disasters are available at: B. Bennett et al., *Health Effects of the Chernobyl Accident and Special Health Care Programmes*, Report of the UN Chernobyl Forum (Geneva: WHO, 2006); World Health

Organization, *Health Risk Assessment from the Nuclear Accident after the 2011 Great East Japan Earthquake and Tsunami Based on a Preliminary Dose Estimation* (Geneva: WHO, 2013).

25 World Nuclear Association, "Nuclear power in France" (accessed 2020), https://www.world-nuclear.org/information-library/country-profiles/countries-a-f/france.aspx.

26 C. Joppke, *Mobilizing Against Nuclear Energy: A Comparison of Germany and the United States* (Berkeley, CA; University of California Press, 1993); Tresantis, *Die Anti-Atom-Bewegung: Geschichte und Perspektiven* (Berlin: Assoziation A, 2015).

27 These points were made repeatedly by Baruch Fischhoff and by Paul Slovic: B. Fischhoff et al., "How safe is safe enough? A psychometric study of attitudes towards technological risks and benefits," *Policy Sciences* 9 (1978), pp. 127–152; B. Fischhoff, "Risk perception and communication unplugged: Twenty years of process," *Risk Analysis* 15/2 (1995), pp. 137–145; B. Fischhoff and J. Kadvany, *Risk: A Very Short Introduction* (New York: Oxford University Press, 2011); P. Slovic, "Perception of risk," *Science* 236/4799 (1987), pp. 280–285; P. Slovic, *The Perception of Risk* (London: Earthscan, 2000); P. Slovic, "Risk perception and risk analysis in a hyperpartisan and virtuously violent world," *Risk Analysis* 40/3 (2020), pp. 2231–2239.

28 Three notable recent disasters indicate the typical range of fatalities during industrial and construction accidents: the derailment, fire, and explosion of a train carrying crude oil at Lac-Mégantic in Quebec (July 6, 2013) with 47 dead; a building collapse in Dhaka killing 1,129 garment workers on April 24, 2013; and the failure of the Brumadinho Dam in Brazil with 233 deaths on January 25, 2019.

29 After free-falling for just four seconds in a belly-to-earth configuration, a base jumper travels 72 m and reaches a speed of 120 km/hour: "BASE jumping freefall chart," *The Great Book of Base* (2010), https://base-book.com/BASEFreefallChart.

30 A. S. Ramírez et al., "Beyond fatalism: Information overload as a mechanism to understand health disparities," *Social Science and Medicine* 219 (2018), pp. 11–18.

31 D. R. Kouabenan, "Occupation, driving experience, and risk and accident perception," *Journal of Risk Research* 5 (2002), pp. 49–68; B. Keeley et al., "Functions of health fatalism: Fatalistic talk as face saving, uncertainty management, stress relief and sense making," *Sociology of Health & Illness* 31 (2009), pp. 734–747.

32 A. Kayani et al., "Fatalism and its implications for risky road use and receptiveness to safety messages: A qualitative investigation in Pakistan," *Health Education Research* 27 (2012), pp. 1043–1054; B. Mahembe and O. M. Samuel, "Influence of personality and fatalistic belief on taxi driver behaviour," *South African Journal of Psychology* 46/3 (2016), pp. 415–426.

33 A. Suárez-Barrientos et al., "Circadian variations of infarct size in acute myocardial infarction," *Heart* 97 (2011), 970e976.

34 World Health Organization, "Falls" (January 2018), https://www.who.int/news-room/fact-sheets/detail/falls.

35 About *Salmonella*, see Centers for Disease Control and Prevention, "*Salmonella* and Eggs", https://www.cdc.gov/foodsafety/communication/salmonella-and-eggs.html. About pesticide residues in tea, see J. Feng et al., "Monitoring and risk assessment of pesticide residues in tea samples from China," *Human and Ecological Risk Assessment: An International Journal* 21/1 (2015), pp. 169–183.

36 The latest FBI statistics for murder and negligent manslaughter (per 100,000 people) are 51 for Baltimore, 9.7 for Miami, and 6.4 for Los Angeles: https://ucr.fbi.gov/crime-in-the-u.s/2018/crime-in-the-u.s.-2018/topic-pages/murder.

37 The largest recent recall of contaminated drugs coming from China included commonly prescribed antihypertensives: Food and Drug Administration, "FDA updates and press announcements on angiotensin II receptor blocker (ARB) recalls (valsartan, losartan, and irbesartan)" (November 2019), https://www.fda.gov/drugs/drug-safety-and-availability/fda-updates-and-press-announcements-angiotensin-ii-receptor-blocker-arb-recalls-valsartan-losartan.

38 Office of National Statistics, "Deaths registered in England and Wales: 2019," https://www.ons.gov.uk/peoplepopulationandcommunity/birthsdeathsandmarriages/deaths/bulletins/deathsregistrationsummarytables/2019.

39 K. D. Kochanek et al., "Deaths: Final Data for 2017," *National Vital Statistics Reports* 68 (2019), pp. 1–75; J. Xu et al., *Mortality in the United States, 2018*, NCHS Data Brief No. 355 (January 2020).

40 Starr, "Social benefit versus technological risk." The micromort metric, introduced in 1989 by Ronald Howard, has been used in many publications by David Spiegelhalter: R. A. Howard, "Microrisks for medical decision analysis," *International Journal of Technology Assessment in Health Care* 5/3 (1989), pp. 357–370; M. Blastland and D. Spiegelhalter, *The Norm Chronicles: Stories and Numbers about Danger and Death* (New York: Basic Books, 2014).

41 United Nations, *World Mortality 2019*, https://www.un.org/en/development/desa/population/publications/pdf/mortality/WMR2019/WorldMortality2019DataBooklet.pdf.

42 CDC, "Heart disease facts," https://www.cdc.gov/heartdisease/facts.htm; D. S. Jones and J. A. Greene, "The decline and rise of coronary heart disease," *Public Health Then and Now* 103 (2014), pp. 10207–10218; J. A. Haagsma et al., "The global burden of injury: incidence, mortality, disability-adjusted life years and time trends from the Global Burden of Disease study 2013," *Injury Prevention* 22/1 (2015), pp. 3–16.

43 World Health Organization, "Falls" (January 2018), https://www.who.int/news-room/fact-sheets/detail/falls.

44 Statistics Canada, "Deaths and mortality rates, by age group" (accessed 2020), https://www150.statcan.gc.ca/t1/tbl1/en/tv.action?pid=1310071001&pickMembers percent5B0 percent5D=1.1&pickMemberspercent5B1 percent5D=3.1.

45 L. T. Kohn et al., *To Err Is Human: Building a Safer Health System* (Washington, DC: National Academies Press, 1999).

46 M. Makary and M. Daniel, "Medical error—the third leading cause of death in the US," *British Medical Journal* 353 (2016), i2139.

47 K. G. Shojania and M. Dixon-Woods, "Estimating deaths due to medical error: the ongoing controversy and why it matters," *British Medical Journal Quality and Safety* 26 (2017), pp. 423–428.

48 J. E. Sunshine et al., "Association of adverse effects of medical treatment with mortality in the United States," *JAMA Network Open* 2/1 (2019), e187041.

49 In 2016, there were 35.7 million hospital stays in the US averaging 4.6 days: W. J. Freeman et al., "Overview of U.S. hospital stays in 2016: Variation by geographic region" (December 2018), https://www.hcup-us.ahrq.gov/reports/statbriefs/sb246-Geographic-Variation-Hospital-Stays.jsp.

50 Bureau of Transportation Statistics, "U.S. Vehicle-miles" (2019), https://www.bts.gov/content/us-vehicle-miles.

51 A. R. Sehgal, "Lifetime risk of death from firearm injuries, drug overdoses, and motor vehicle accidents in the United States," *American Journal of Medicine* 133/10 (October 2020), pp. 1162–1167.

52 World Health Rankings, "Road traffic accidents" (accessed 2020), https://www.worldlifeexpectancy.com/cause-of-death/road-traffic-accidents/by-country/.

53 The mystery of Malaysia Airlines flight 370 may never be solved: suggestions and speculations abound but at this time it seems that only some unexpected, accidental key may unlock it. Investigation of the two successive crashes of Boeing 737 MAX (killing 346 people) exposed the company's questionable practices in manufacturing its bestselling design and in offering instructions and guidance for its operation.

54 International Civil Aviation Organization, *State of Global Aviation Safety* (Montreal: ICAO, 2020).

55 K. Soreide et al., "How dangerous is BASE jumping? An analysis of adverse events in 20,850 jumps from the Kjerag Massif, Norway," *Trauma* 62/5 (2007), pp. 1113–1117.

56 United States Parachute Association, "Skydiving safety" (accessed 2020), https://uspa.org/Find/FAQs/Safety.

57 US Hang Gliding & Paragliding Association, "Fatalities" (accessed 2020), https://www.ushpa.org/page/fatalities.

58 National Consortium for the Study of Terrorism and Responses to Terrorism, *American Deaths in Terrorist Attacks, 1995–2017* (September 2018).

59 National Consortium for the Study of Terrorism and Responses to Terrorism, *Trends in Global Terrorism: Islamic State's Decline in Iraq and Expanding Global Impact; Fewer Mass Casualty Attacks in Western Europe; Number of Attacks in the United States Highest since 1980s* (October 2019).

60 For a good summary of the West Coast earthquake perils, see R. S. Yeats, *Living with Earthquakes in California* (Corvallis, OR: Oregon State University Press, 2001). For the transpacific consequences of the West Coast earthquakes, see B. F. Atwater, *The Orphan Tsunami of 1700* (Seattle, WA: University of Washington Press, 2005).

61 E. Agee and L. Taylor, "Historical analysis of U.S. tornado fatalities (1808–2017): Population, science, and technology," *Weather, Climate and Society* 11 (2019), pp. 355–368.

62 R. J. Samuels, *3.11: Disaster and Change in Japan* (Ithaca, NY: Cornell University Press, 2013); V. Santiago-Fandiño et al., eds., *The 2011 Japan Earthquake and Tsunami: Reconstruction and Restoration, Insights and Assessment after 5 Years* (Berlin: Springer, 2018).

63 E. N. Rappaport, "Fatalities in the United States from Atlantic tropical cyclones: New data and interpretation," *Bulletin of American Meteorological Society* 1014 (March 2014), pp. 341–346.

64 National Weather Service, "How dangerous is lightning?" (accessed 2020), https://www.weather.gov/safety/lightning-odds; R. L. Holle et al., "Seasonal, monthly, and weekly distributions of NLDN and GLD360 cloud-to-ground lightning," *Monthly Weather Review* 144 (2016), pp. 2855–2870.

65 Munich Re, *Topics. Annual Review: Natural Catastrophes 2002* (Munich: Munich Re, 2003); P. Löw, "Tropical cyclones cause highest losses: Natural disasters of 2019 in figures," Munich Re (January 2020), https://www.munichre.com/topics-online/en/climate-change-and-natural-disasters/natural-disasters/natural-disasters-of-2019-in-figures-tropical-cyclones-cause-highest-losses.html.

66 O. Unsalan et al., "Earliest evidence of a death and injury by a meteorite," *Meteoritics & Planetary Science* (2020), pp. 1–9.

67 National Research Council, *Near-Earth Object Surveys and Hazard Mitigation Strategies: Interim Report* (Washington, DC: NRC, 2009); M. A. R. Khan, "Meteorites," *Nature* 136/1030 (1935), p. 607.

68 D. Finkelman, "The dilemma of space debris," *American Scientist* 102/1 (2014), pp. 26–33.

69 M. Mobberley, *Supernovae and How to Observe Them* (New York: Springer, 2007).

70 NASA, "2012: Fear no Supernova" (December 2011), https://www.nasa.gov/topics/earth/features/2012-supernova.html.

71 NASA, "Asteroid fast facts" (March 2014), https://www.nasa.gov/mission_pages/asteroids/overview/fastfacts.html; National Research Council, *Near-Earth Object Surveys and Hazard Mitigation Strategies*; M. B. E. Boslough and D. A. Crawford, "Low-altitude airbursts and the impact threat," *International Journal of Impact Engineering* 35/12 (2008), pp. 1441–1448.

72 US Geological Survey, "What would happen if a 'supervolcano' eruption occurred again at Yellowstone?" https://www.usgs.gov/faqs/what-would-happen-if-a-supervolcano-eruption-occurred-again-yellowstone; R. V. Fisher et al., *Volcanoes: Crucibles of Change* (Princeton, NJ: Princeton University Press, 1997).

73 Space Weather Prediction Center, "Coronal mass ejections," National Oceanic and Atmospheric Administration (accessed 2020), https://www.swpc.noaa.gov/phenomena/coronal-mass-ejections.

74 R. R. Britt, "150 years ago: The worst solar storm ever," Space.com (September 2009), https://www.space.com/7224-150-years-worst-solar-storm.html.

75 S. Odenwald, "The day the Sun brought darkness," NASA (March 2009), https://www.nasa.gov/topics/earth/features/sun_darkness.html.

76 Solar and Heliospheric Observatory, https://sohowww.nascom.nasa.gov/.

77 T. Phillips, "Near miss: The solar superstorm of July 2012," NASA (July 2014), https://science.nasa.gov/science-news/science-at-nasa/2014/23jul_superstorm.

78 P. Riley, "On the probability of occurrence of extreme space weather events," *Space Weather* 10 (2012), S02012.

79 D. Moriña et al., "Probability estimation of a Carrington-like geomagnetic storm," *Scientific Reports* 9/1 (2019).

80 K. Kirchen et al., "A solar-centric approach to improving estimates of exposure processes for coronal mass ejections," *Risk Analysis* 40 (2020), pp. 1020–1039.

81 E. D. Kilbourne, "Influenza pandemics of the 20th century," *Emerging Infectious Diseases* 12/1 (2006), pp. 9–14.

82 C. Viboud et al., "Global mortality impact of the 1957–1959 influenza pandemic," *Journal of Infectious Diseases* 213/5 (2016), pp. 738–745; CDC, "1968 Pandemic (H3N2 virus)" (accessed 2020), https://www.cdc.gov/flu/pandemic-resources/1968-pandemic.html; J. Y. Wong et al., "Case fatality risk of influenza A(H1N1pdm09): a systematic review," *Epidemiology* 24/6 (2013).

83 World Economic Forum, *Global Risks 2015, 10th Edition* (Cologny: WEF, 2015).

84 "Advice on the use of masks in the context of COVID-19: Interim guidance," World Health Organization (2020).

85 J. Paget et al., "Global mortality associated with seasonal influenza epidemics: New burden estimates and predictors from the GLaMOR Project," *Journal of Global Health* 9/2 (December 2019), 020421.

86 W. Yang et al., "The 1918 influenza pandemic in New York City: Age-specific timing, mortality, and transmission dynamics," *Influenza and Other Respiratory Viruses* 8 (2014), pp. 177–188; A. Gagnon et al., "Age-specific mortality during the 1918 influenza pandemic: Unravelling the mystery of high young adult mortality," *PLoS ONE* 8/8 (August 2013), e6958; W. Gua et al., "Comorbidity and its impact on 1590 patients with COVID-19 in China: A nationwide analysis," *European Respiratory Journal* 55/6 (2020), article 2000547.

87 J.-M. Robine et al., eds., *Human Longevity, Individual Life Duration, and the Growth of the Oldest-Old Population* (Berlin: Springer, 2007).

88 CDC, "Weekly Updates by Select Demographic and Geographic Characteristics" (accessed 2020), https://www.cdc.gov/nchs/nvss/vsrr/covid_weekly/index.htm#AgeAndSex.

89 D. M. Morens et al., "Predominant role of bacterial pneumonia as a cause of death in pandemic influenza: implications for pandemic influenza preparedness," *Journal of Infectious Disease* 198/7 (October 2008), pp. 962–970.

90 A. Noymer and M. Garenne, "The 1918 influenza epidemic's effects on sex differentials in mortality in the United States," *Population and Development Review* 26/3 (2000), pp. 565–581.

91 For a good summary of the West Coast earthquake perils, see R. S. Yeats, *Living with Earthquakes in California* (Corvallis, OR: Oregon

State University Press, 2001). For the transpacific consequences of the West Coast earthquakes, see B. F. Atwater, *The Orphan Tsunami of 1700* (Seattle, WA: University of Washington Press, 2005).

92 P. Gilbert, *The A-Z Reference Book of Syndromes and Inherited Disorders* (Berlin: Springer, 1996).

93 Japan, with its population concentrated in the lowlands that make up only about 15 percent of this mountainous country, and with its ever-present risk of powerful earthquakes, volcanic eruptions, and destructive tsunami, is a prime example of this reality—as are, for similar and other reasons, such densely populated places as Java or coastal Bangladesh.

94 Much more on these topics can be found in many recent publications, including O. Renn, *Risk Governance: Towards an Integrative Approach* (Geneva: International Risk Governance Council, 2006); G. Gigerenzer, *Risk Savvy: How to Make Good Decisions* (New York: Penguin Random House, 2015).

95 V. Janssen, "When polio triggered fear and panic among parents in the 1950s," *History* (March 2020), https://www.history.com/news/polio-fear-post-wwii-era.

96 In 1958, the US GDP increased by more than 5 percent above the 1957 level, and the gain was more than 7 percent in 1969. Fred Economic Data (accessed 2020), https://fred.stlouisfed.org/series/GDP.

97 The Museum of Flight, "Boeing 747-121" (accessed 2020), https://www.museumofflight.org/aircraft/boeing-747-121.

98 Y. Tsuji et al., "Tsunami heights along the Pacific Coast of Northern Honshu recorded from the 2011 Tohoku and previous great earthquakes," *Pure and Applied Geophysics* 171 (2014), pp. 3183–3215.

99 In November 2004, to American citizens, Osama bin Laden explained that he chose that attack in order to bleed "America to the point of bankruptcy," and how this was helped by "the White House that demands the opening of war fronts." A full transcript of the speech is available at: https://www.aljazeera.com/archive/2004/11/200849163336457223.html. He also cited the Royal Institute of International Affairs' estimate that mounting the attacks

cost no more than $500,000, while by 2018 the cost of America's wars in Iraq, Afghanistan, Pakistan, and Syria rose to about $5.9 trillion, and future costs (interest on borrowed monies, veteran care) could push this to $8 trillion in the next 40 years: Watson Institute, "Costs of War" (2018), https://watson.brown.edu/costsofwar/papers/summary.

100 C. R. Sunstein, "Terrorism and probability neglect," *Journal of Risk and Uncertainty* 26 (2003), pp. 121–136.

101 Federal Bureau of Investigation, "Crime in the U.S." (accessed 2020), https://ucr.fbi.gov/crime-in-the-u.s.

102 E. Miller and N. Jensen, *American Deaths in Terrorist Attacks, 1995–2017* (September 2018), https://www.start.umd.edu/pubs/START_AmericanTerrorismDeaths_FactSheet_Sept2018.pdf.

103 A. R. Sehgal, "Lifetime risk of death from firearm injuries, drug overdoses, and motor vehicle accidents in the United States," *American Journal of Medicine* 133/10 (May 2020), pp. 1162–1167.

6. Understanding the Environment: The Only Biosphere We Have

1 For the most delusionary version of these visions, see https://www.spacex.com/mars. Their self-imposed milestones: The first mission to Mars in 2022; its modest goals are "to confirm water resources, identify hazards, and put in place initial power, mining, and life support infrastructure." The second mission, in 2024, will build a propellant depot, prepare for future crew flights, and "serve as the beginnings of the first Mars base, from which we can build a thriving city and eventually self-sustaining civilization on Mars." Those who enjoy this fantasy genre may also check: K. M. Cannon and D. T. Britt, "Feeding one million people on Mars," *New Space* 7/4 (December 2019), pp. 245–254.

2 B. M. Jakosky and C. S. Edwards, "Inventory of CO_2 available for terraforming Mars," *Nature Astronomy* 2 (2018), pp. 634–639.

3 This was discussed by a webinar hosted by the New York Academy of Sciences in May 2020, when a geneticist from Cornell University even said "And are we maybe ethically bound to do

so?": "Alienating Mars: Challenges of Space Colonization," https://www.nyas.org/events/2020/webinar-alienating-mars-challenges-of-space-colonization. Remarkably, this vision of people endowed with tardigrade-like genetic resilience was discussed, apparently in all seriousness, at a time when New York City was recording more than 500 deaths daily from COVID-19 and when hospitals faced constant shortages of simple PPE and were forced to reuse masks and gloves. The Defense Advanced Research Project Agency has been also spending public money on this: J. Koebler, "DARPA: We Are Engineering the Organism that will Terraform Mars," VICE Motherboard (June 2015), https://www.vice.com/en_us/article/ae3pee/darpa-we-are-engineering-the-organisms-that-will-terraform-mars.

4 J. Rockström et al., "A safe operating space for humanity," *Nature* 461 (2009), pp. 472–475.

5 For complete lists of all categories of freediving and static apnea records, see https://www.guinnessworldrecords.com/search?term=freediving.

6 Average tidal volume (air intake into lungs) is 500 mL for men and 400 mL for women: S. Hallett and J. V. Ashurst, "Physiology, tidal volume" (June 2020), https://www.ncbi.nlm.nih.gov/books/NBK482502/. Taking 450 mL and 16 intakes per minute as the means, this makes 7.2 liters of air per minute. Oxygen makes up almost 21 percent of the air and hence about 1.5 liters of it are inhaled every minute, but only about 23 percent of that volume is absorbed by lungs (the rest is expired) and the actual consumption of pure oxygen is about 350 mL per minute—that is, 500 L or (with 1.429 g/L) about 700 grams a day. Physical exertion increases the need, and with just 30 percent markup for higher oxygen consumption during daily activities this comes to about 900 g a day. For maximum oxygen intakes, see G. Ferretti, "Maximal oxygen consumption in healthy humans: Theories and facts," *European Journal of Applied Physiology* 114 (2014), pp. 2007–2036.

7 A. P. Gumsley et al., "Timing and tempo of the Great Oxidation Event," *Proceedings of the National Academy of Sciences* 114 (2017), pp. 1811–1816.

8 R. A. Berner, "Atmospheric oxygen over Phanerozoic time," *Proceedings of the National Academy of Sciences* 96 (1999), pp. 10955–10957.
9 For carbon content of terrestrial vegetation, see V. Smil, *Harvesting the Biosphere* (Cambridge, MA: MIT Press, 2013), pp. 161–165. The calculation assumes complete oxidation of all of this carbon.
10 https://twitter.com/EmmanuelMacron/status/1164617008962527232.
11 S. A. Loer et al., "How much oxygen does the human lung consume?" *Anesthesiology* 86 (1997), pp. 532–537.
12 Smil, *Harvesting the Biosphere*, pp. 31–36.
13 J. Huang et al., "The global oxygen budget and its future projection," *Science Bulletin* 63/18 (2018), pp. 1180–1186.
14 There are, of course, many other real reasons—ranging from the loss of biodiversity to changes in water retention capacity—to be concerned about the deliberate large-scale burning of tropical vegetation or about natural fires in drought-stricken forests.
15 For the latest surveys of global water supply and use, see A. K. Biswas et al., eds., *Assessing Global Water Megatrends* (Singapore: Springer Nature, 2018).
16 Institute of Medicine, *Dietary Reference Intakes for Water, Potassium, Sodium, Chloride, and Sulfate* (Washington, DC: National Academies Press, 2005).
17 Among the world's most populous nations, agriculture's share of freshwater withdrawals is as high as 90 percent in India, 80 percent in Indonesia, and 65 percent in China, but only about 35 percent in the United States: World Bank, "Annual freshwater withdrawals, agriculture (percent of total freshwater withdrawal)" (accessed 2020), https://data.worldbank.org/indicator/er.h2o.fwag.zs?end=2016&start=1965&view=chart.
18 Water Footprint Network, "What is a water footprint?" (accessed 2020), https://waterfootprint.org/en/water-footprint/what-is-water-footprint/.
19 M. M. Mekonnen and Y. A. Hoekstra, *National Water Footprint Accounts: The Green, Blue and Grey Water Footprint of Production and Consumption* (Delft: UNESCO-IHE Institute for Water Education, 2011).

20 N. Joseph et al., "A review of the assessment of sustainable water use at continental-to-global scale," *Sustainable Water Resources Management* 6 (2020), p. 18.
21 S. N. Gosling and N.W. Arnell, "A global assessment of the impact of climate change on water scarcity," *Climatic Change* 134 (2016), pp. 371–385.
22 Smil, *Growth*, pp. 386–388.
23 For the long-term trends of different categories of agricultural land use, see FAO, "Land use," http://www.fao.org/faostat/en/#data/RL. An American study put 2009 as the year of the global peak in farmland, followed by a steady slow decline: J. Ausubel et al., "Peak farmland and the prospect for land sparing," *Population and Development Review* 38, Supplement (2012), pp. 221–242. In reality, FAO data show another 4 percent increase between 2009 and 2017.
24 X. Chen et al., "Producing more grain with lower environmental costs," *Nature* 514/7523 (2014), pp. 486–488; Z. Cui et al., "Pursuing sustainable productivity with millions of smallholder farmers," *Nature* 555/7696 (2018), pp. 363–366.
25 Global output of ammonia contained 160 Mt of nitrogen in 2019, with about 120 destined for fertilizers: FAO, *World Fertilizer Trends and Outlook to 2022* (Rome: FAO, 2019). Production capacity (already in excess of 180 Mt) is expected to increase by nearly 20 percent by 2026, with some 100 planned and announced plants, primarily in Asia and the Middle East: Hydrocarbons Technology, "Asia and Middle East lead globally on ammonia capacity additions" (2018), https://www.hydrocarbons-technology.com/comment/global-ammonia-capacity/.
26 US Geological Survey, "Potash" (2020), https://pubs.usgs.gov/periodicals/mcs2020/mcs2020-potash.pdf.
27 J. Grantham, "Be persuasive. Be brave. Be arrested (if necessary)," *Nature* 491 (2012), p. 303.
28 S. J. Van Kauwenbergh, *World Phosphate Rock Reserves and Resources* (Muscle Shoals, AL: IFDC, 2010).
29 US Geological Survey, *Mineral Commodity Summaries 2012*, p. 123.
30 International Fertilizer Industry Association, "Phosphorus and 'Peak Phosphate'" (2013). See also M. Heckenmüller et al., *Global Availability*

of Phosphorus and Its Implications for Global Food Supply: An Economic Overview (Kiel: Kiel Institute for the World Economy, 2014).

31 V. Smil, "Phosphorus in the environment: Natural flows and human interferences," *Annual Review of Energy and the Environment* 25 (2000), pp. 53–88; US Geological Survey, "Phosphate rock," https://pubs.usgs.gov/periodicals/mcs2020/mcs2020-phosphate.pdf.

32 M. F. Chislock et al., "Eutrophication: Causes, consequences, and controls in aquatic ecosystems," *Nature Education Knowledge* 4/4 (2013), p. 10.

33 J. Bunce et al., "A review of phosphorus removal technologies and their applicability to small-scale domestic wastewater treatment systems," *Frontiers in Environmental Science* 6 (2018), p. 8.

34 D. Breitburg et al., "Declining oxygen in the global ocean and coastal waters," *Science* 359/6371 (2018).

35 R. Lindsey, "Climate and Earth's energy budget," NASA (January 2009), https://earthobservatory.nasa.gov/features/EnergyBalance.

36 W. F. Ruddiman, *Plows, Plagues & Petroleum: How Humans Took Control of Climate* (Princeton, NJ: Princeton University Press, 2005).

37 2° Institute, "Global CO_2 levels" (accessed 2020), https://www.co2levels.org/.

38 2° Institute, "Global CH_4 levels" (accessed 2020), https://www.methanelevels.org/.

39 Global warming potentials (CO_2=1) are 28 for methane, 265 for nitrous dioxide, 5,660 to 13,900 for various chlorofluorocarbons, and 23,900 for sulfur hexafluoride: Global Warming Potential Values, https://www.ghgprotocol.org/sites/default/files/ghgp/Global-Warming-Potential-Values%20%28Feb%2016%202016%29_1.pdf.

40 IPCC, *Climate Change 2014: Synthesis Report. Contribution of Working Groups I, II and III to the Fifth Assessment Report of the Intergovernmental Panel on Climate Change* (Geneva: IPCC, 2014).

41 J. Fourier, "Remarques générales sur les Temperatures du globe terrestre et des espaces planetaires," *Annales de Chimie et de*

Physique 27 (1824), pp. 136–167; E. Foote, "Circumstances affecting the heat of the sun's rays," *American Journal of Science and Arts* 31 (1856), pp. 382–383. Foote's clear conclusion: "The highest effect of the sun's rays I have found to be in carbonic acid gas . . . An atmosphere of that gas would give to our earth a high temperature; and if as some suppose, at one period of its history the air had mixed with it a larger proportion than at present, an increased temperature from its own action as well as from increased weight must have necessarily resulted".

42 J. Tyndall, "The Bakerian Lecture," *Philosophical Transactions* 151 (1861), pp. 1–37 (quote p. 28).

43 S. Arrhenius, "On the influence of carbonic acid in the air upon the temperature of the ground," *Philosophical Magazine and Journal of Science*, 5/41 (1896), pp. 237–276.

44 K. Ecochard, "What's causing the poles to warm faster than the rest of the Earth?" NASA (April 2011), https://www.nasa.gov/topics/earth/features/warmingpoles.html.

45 D. T. C. Cox et al., "Global variation in diurnal asymmetry in temperature, cloud cover, specific humidity and precipitation and its association with leaf area index," *Global Change Biology* (2020).

46 S. Arrhenius, *Worlds in the Making* (New York: Harper & Brothers, 1908), p. 53.

47 R. Revelle and H. E. Suess, "Carbon dioxide exchange between atmosphere and ocean and the question of an increase of atmospheric CO_2 during the past decades," *Tellus* 9 (1957), pp. 18–27.

48 Global Monitoring Laboratory, "Monthly average Mauna Loa CO_2" (accessed 2020), https://www.esrl.noaa.gov/gmd/ccgg/trends/.

49 J. Charney et al., *Carbon Dioxide and Climate: A Scientific Assessment* (Washington, DC: National Research Council, 1979).

50 N. L. Bindoff et al., "Detection and Attribution of Climate Change: from Global to Regional," in T. F. Stocker et al., eds., *Climate Change 2013: The Physical Science Basis. Contribution of Working Group I to the Fifth Assessment Report of the Intergovernmental Panel on Climate Change* (Cambridge: Cambridge University Press, 2013).

51 S. C. Sherwood et al., "An assessment of Earth's climate sensitivity using multiple lines of evidence," *Reviews of Geophysics* 58/4 (December 2020).

52 Switching from coal to natural gas has been remarkably rapid in the US: in 2011, 44 percent of all electricity was generated by coal; by 2020 that share fell to just 20 percent; while gas-fired generation rose from 23 percent to 39 percent: US EIA, *Short-Term Energy Outlook* (2021).

53 In 2014, the global mean of anthropogenic forcing relative to 1850 was 1.97 W/m^2, with 1.80 W/m^2 from CO_2, 1.07 W/m^2 from other well-mixed greenhouse gases, −1.04 W/m^2 from aerosols, and −0.08 W/m^2 from land use changes: C. J. Smith et al., "Effective radiative forcing and adjustments in CMIP6 models," *Atmospheric Chemistry and Physics* 20/16 (2020).

54 National Centers for Environmental Information, "More near-record warm years are likely on the horizon" (February 2020), https://www.ncei.noaa.gov/news/projected-ranks; NOAA, *Global Climate Report—Annual 2019*, https://www.ncdc.noaa.gov/sotc/global/201913.

55 For Kyoto cherries, see: R. B. Primack et al., "The impact of climate change on cherry trees and other species in Japan," *Biological Conservation* 142 (2009), pp. 1943–1949. For the French vintages, see Ministère de la Transition Écologique, "Impacts du changement climatique: Agriculture et Forêt" (2020), https://www.ecologie.gouv.fr/impacts-du-changement-climatique-agriculture-et-foret. For the melting of mountain glaciers and its consequences, see A. M. Milner et al., "Glacier shrinkage driving global changes in downstream systems," *Proceedings of the National Academy of Sciences* (2017), www.pnas.org/cgi/doi/10.1073/pnas.1619807114.

56 In 2019, fossil fuel combustion released nearly 37 Gt of CO_2, whose generation required very close to 27 Gt of oxygen: Global Carbon Project, *The Global Carbon Budget 2019*.

57 J. Huang et al., "The global oxygen budget and its future projection," *Science Bulletin* 63 (2018), pp. 1180–1186.

58 These intricate measurements began in 1989: Carbon Dioxide Information and Analysis Center, "Modern Records of Atmospheric

Oxygen (O_2) from Scripps Institution of Oceanography" (2014), https://cdiac.ess-dive.lbl.gov/trends/oxygen/modern_records.html.

59 Reserves of fossil fuels for 2019 are listed in British Petroleum, *Statistical Review of World Energy*.

60 L. B. Scheinfeldt and S. A. Tishkoff, "Living the high life: high-altitude adaptation," *Genome Biology* 11/133 (2010), pp. 1–3.

61 S. J. Murray et al., "Future global water resources with respect to climate change and water withdrawals as estimated by a dynamic global vegetation model," *Journal of Hydrology* (2012), pp. 448–449; A. G. Koutroulis and L. V. Papadimitriou, "Global water availability under high-end climate change: A vulnerability based assessment," *Global and Planetary Change* 175 (2019), pp. 52–63.

62 P. Greve et al., "Global assessment of water challenges under uncertainty in water scarcity projections," *Nature Sustainability* 1/9 (2018), pp. 486–494.

63 C. A. Dieter et al., *Estimated Use of Water in the United States in 2015* (Washington, DC: US Geological Survey, 2018).

64 P. S. Goh et al., *Desalination Technology and Advancement* (Oxford: Oxford Research Encyclopedias, 2019).

65 A. Fletcher et al., "A low-cost method to rapidly and accurately screen for transpiration efficiency in wheat," *Plant Methods* 14 (2018), article 77. Whole-plant transpiration efficiency of 4.5 g/kg means that 1 kg of biomass requires 222 kg of transpired water, and with grain being about half of the total aboveground biomass, the ratio doubles to nearly 450 kg.

66 Y. Markonis et al., "Assessment of water cycle intensification over land using a multisource global gridded precipitation dataset," *Journal of Geophysical Research: Atmospheres* 124/21 (2019), pp. 11175–11187.

67 S. J. Murray et al., "Future global water resources with respect to climate change and water withdrawals as estimated by a dynamic global vegetation model."

68 Y. Fan et al., "Comparative evaluation of crop water use efficiency, economic analysis and net household profit simulation in arid Northwest China," *Agricultural Water Management* 146 (2014), pp. 335–345; J. L. Hatfield and C. Dold, "Water-use efficiency:

Advances and challenges in a changing climate," *Frontiers in Plant Science* 10 (2019), p. 103; D. Deryng et al., "Regional disparities in the beneficial effects of rising CO_2 concentrations on crop water productivity," *Nature Climate Change* 6 (2016), pp. 786–790.

69 IPCC, *Climate Change and Land* (Geneva: IPCC, 2020), https://www.ipcc.ch/srccl/; P. Smith et al., "Agriculture, Forestry and Other Land Use (AFOLU)," in IPCC, *Climate Change 2014*.

70 Smil, *Should We Eat Meat?*, pp. 203–210.

71 D. Gerten et al., "Feeding ten billion people is possible within four terrestrial planetary boundaries," *Nature Sustainability* 3 (2020), pp. 200–208; see also FAO, *The Future of Food and Agriculture: Alternative Pathways to 2050* (Rome: FAO, 2018), http://www.fao.org/3/I8429EN/i8429en.pdf.

72 I wrote: "Adding the mean and the highest interval [between the successive pandemics] to 1968 gives a span between 1996 and 2021. We are, probabilistically speaking, very much inside a high-risk zone. Consequently, the likelihood of another influenza pandemic during the next 50 years is virtually 100 percent": V. Smil, *Global Catastrophes and Trends* (Cambridge, MA: MIT Press, 2008), p. 46. And we got two pandemics within the indicated interval: H1N1 virus in 2009, the year after the book was published, and SARS-Cov-2 in 2020.

73 Global daily statistical updates were provided by Johns Hopkins at https://coronavirus.jhu.edu/map.html and by Worldometer at https://www.worldometers.info/coronavirus/. We will have to wait, at least for two years, for a truly comprehensive history of the pandemic.

74 U. Desideri and F. Asdrubali, *Handbook of Energy Efficiency in Buildings* (London: Butterworth-Heinemann, 2015).

75 Natural Resource Canada, *High Performance Housing Guide for Southern Manitoba* (Ottawa: Natural Resources Canada, 2016).

76 L. Cozzi and A. Petropoulos, "Growing preference for SUVs challenges emissions reductions in passenger car market," IEA (October 2019), https://www.iea.org/commentaries/growing-preference-for-suvs-challenges-emissions-reductions-in-passenger-car-market.

77 J. G. J. Olivier and J. A. H. W. Peters, *Trends in Global CO_2 and Total Greenhouse Gas Emissions* (The Hague: PBL Netherlands Environmental Assessment Agency, 2019).

78 United Nations, "Conference of the Parties (COP), https://unfccc.int/process/bodies/supreme-bodies/conference-of-the-parties-cop.

79 N. Stockton, "The Paris climate talks will emit 300,000 tons of CO_2, by our math. Hope it's worth it," *Wired* (November 2015).

80 United Nations, *Report of the Conference of the Parties on its twenty-first session, held in Paris from 30 November to 13 December 2015* (January 2016), https://unfccc.int/sites/default/files/resource/docs/2015/cop21/eng/10a01.pdf.

81 For the future of air conditioning, see International Energy Agency, *The Future of Cooling* (Paris: IEA, 2018).

82 Olivier and Peters, *Trends in Global CO_2 and Total Greenhouse Gas Emissions* 2019 Report.

83 T. Mauritsen and R. Pincus, "Committed warming inferred from observations," *Nature Climate Change* 7 (2017), pp. 652–655.

84 C. Zhou et al., "Greater committed warming after accounting for the pattern effect," *Nature Climate Change* 11 (2021), pp. 132–136.

85 IPCC, *Global warming of 1.5°C* (Geneva: IPCC, 2018), https://www.ipcc.ch/sr15/.

86 A. Grubler et al., "A low energy demand scenario for meeting the 1.5°C target and sustainable development goals without negative emission technologies," *Nature Energy* 526 (2020), pp. 515–527.

87 European Environment Agency, "Size of the vehicle fleet in Europe" (2019), https://www.eea.europa.eu/data-and-maps/indicators/size-of-the-vehicle-fleet/size-of-the-vehicle-fleet-10; for 1990, see https://www.eea.europa.eu/data-and-maps/indicators/access-to-transport-services/vehicle-ownership-term-2001.

88 National Bureau of Statistics, *China Statistical Yearbook, 1999-2019*, http://www.stats.gov.cn/english/Statisticaldata/AnnualData/.

89 SEI, IISD, ODI, E3G, and UNEP, *The Production Gap Report: 2020 Special Report*, http://productiongap.org/2020report.

90 E. Larson et al., *Net-Zero America: Potential Pathways, Infrastructure, and Impacts* (Princeton, NJ: Princeton University, 2020).

91 C. Helman, "Nimby nation: The high cost to America of saying no to everything," *Forbes* (August 2015).

92 The House of Representatives, "Resolution Recognizing the duty of the Federal Government to create a Green New Deal" (2019), https://www.congress.gov/bill/116th-congress/house-resolution/109/text; M. Z. Jacobson et al., "Impacts of Green New Deal energy plans on grid stability, costs, jobs, health, and climate in 143 countries," *One Earth* 1 (2019), pp. 449–463.

93 T. Dickinson, "The Green New Deal is cheap, actually," *Rolling Stone* (April 6, 2020); J. Cassidy, "The good news about a Green New Deal," *New Yorker* (March 4, 2019); N. Chomsky and R. Pollin, *Climate Crisis and the Global Green New Deal: The Political Economy of Saving the Planet* (New York: Verso, 2020); J. Rifkin, *The Green New Deal: Why the Fossil Fuel Civilization Will Collapse by 2028, and the Bold Economic Plan to Save Life on Earth* (New York: St. Martin's Press, 2019).

94 If you want to join the most explicit branch of this movement—in order to mobilize "3.5 percent of the population to achieve system change" (rebellion down to a decimal point!)—check: Extinction Rebellion, "Welcome to the rebellion," https://rebellion.earth/the-truth/about-us/. For written instructions, see Extinction Rebellion, *This Is Not a Drill: An Extinction Rebellion Handbook* (London: Penguin, 2019).

95 P. Brimblecombe et al., *Acid Rain—Deposition to Recovery* (Berlin: Springer, 2007).

96 S. A. Abbasi and T. Abbasi, *Ozone Hole: Past, Present, Future* (Berlin: Springer, 2017).

97 J. Liu et al., "China's changing landscape during the 1990s: Large-scale land transformation estimated with satellite data," *Geophysical Research Letters* 32/2 (2005), L02405.

98 M. G. Burgess et al., "IPCC baseline scenarios have over-projected CO_2 emissions and economic growth," *Environmental Research Letters* 16 (2021), 014016.

99 H. Wood, "Green energy meets people power," *The Economist* (2020), https://worldin.economist.com/article/17505/edition2020 get-ready-renewable-energy-revolution.

100 Z. Hausfather et al., "Evaluating the performance of past climate model projections," *Geophysical Research Letters* 47 (2019), e2019 GL085378.

101 Smil, "History and risk."

102 Global and national daily and cumulative totals from Johns Hopkins at https://coronavirus.jhu.edu/map.html or from Worldometer at https://www.worldometers.info/coronavirus/.

103 Sources for data in this, and the following, paragraph are as follows. For GDP rate, see World Bank, "GDP per capita (current US$)" (accessed 2020), https://data.worldbank.org/indicator/NY.GDP.PCAP.CD. For the Chinese statistics, see National Bureau of Statistics, *China Statistical Yearbook, 1999–2019*. For national CO_2 emissions, see Olivier and Peters, *Trends in Global CO_2 and Total Greenhouse Gas Emissions* 2019 Report.

104 Between 2020 and 2050, the UN's medium population forecast projects 99.6 percent of the total increase taking place in less developed countries and about 53 percent of the total in sub-Saharan Africa: United Nations, *World Population Prospects: The 2019 Revision* (New York: UN, 2019). On Africa's electricity generation lock-in, see G. Alova et al., "A machine-learning approach to predicting Africa's electricity mix based on planned power plants and their chances of success," *Nature Energy* 6/2 (2021).

105 Y. Pan et al., "Large and persistent carbon sink in the world's forests," *Science* 333 (2011), pp. 988–993; C. Che et al., "China and India lead in greening of the world through land-use management," *Nature Sustainability* 2 (2019), pp. 122–129. See also J. Wang et al., "Large Chinese land carbon sink estimated from atmospheric carbon dioxide data," *Nature* 586/7831 (2020), pp. 720–723.

106 N. G. Dowell et al., "Pervasive shifts in forest dynamics in a changing world," *Science* 368 (2020); R. J. W. Brienen et al., "Forest carbon sink neutralized by pervasive growth-lifespan trade-offs," *Nature Communications* 11 (2020), article 4241.12345 67890.

107 P. E. Kauppi et al., "Changing stock of biomass carbon in a boreal forest over 93 years," *Forest Ecology and Management* 259 (2010), pp. 1239–1244; H. M. Henttonen et al., "Size-class

structure of the forests of Finland during 1921–2013: A recovery from centuries of exploitation, guided by forest policies," *European Journal of Forest Research* 139 (2019), pp. 279–293.

108 P. Roy and J. Connell, "Climatic change and the future of atoll states," *Journal of Coastal Research* 7 (1991), pp. 1057–1075; R. J. Nicholls and A. Cazenave, "Sea-level rise and its impact on coastal zones," *Science* 328/5985 (2010), pp. 1517–1520.

109 P. S. Kench et al., "Patterns of island change and persistence offer alternate adaptation pathways for atoll nations," *Nature Communications* 9 (2018), article 605.

110 This was the title of a chapter contributed by Amory Lovins to a book on the global environment: A. Lovins, "Abating global warming for fun and profit," in K. Takeuchi and M. Yoshino, eds., *The Global Environment* (New York: Springer-Verlag, 1991), pp. 214–229. For the younger readers: Lovins established his fame with a 1976 paper where he outlined the "soft" (small-scale renewable) energy path for the US: A. Lovins, "Energy strategy: The road not taken," *Foreign Affairs* 55/1 (1976), pp. 65–96. According to his vision, in the year 2000 the US was to derive energy equal to about 750 million tons of oil equivalent from soft techniques. After subtracting conventional large-scale hydrogeneration (neither small nor soft), the renewables contributed the equivalent of just over 75 million tons of oil, and Lovins thus missed his target by 90 percent in 24 years, a forecast that presaged decades of similarly unrealistic "green" claims.

7. Understanding the Future: Between Apocalypse and Singularity

1 Books on apocalypticism and apocalyptic prophecies, imagination, and interpretations are quite numerous, but I would not presume to make any recommendations regarding this particular kind of fiction writing.

2 To imagine that artificial intelligence will surpass human capability is easy compared to imagining an instantaneous rate of physical change that is required by reaching the Singularity.

3 R. Kurzweil, "The law of accelerating returns" (2001), https://www.kurzweilai.net/the-law-of-accelerating-returns. See also his *The Singularity Is Near* (New York: Penguin, 2005). The 2045 arrival is predicted in https://www.kurzweilai.net/. Before we get there, "by the 2020s, most diseases will go away as nanobots become smarter than current medical technology. Normal human eating can be replaced by nanosystems." See P. Diamandis, "Ray Kurzweil's mind-boggling predictions for the next 25 years," Singularity Hub (January 2015), https://singularityhub.com/2015/01/26/ray-kurzweils-mind-boggling-predictions-for-the-next-25-years/. Obviously, if such forecasts were to materialize, then in just a few years nobody would have to write books about agriculture, food, health, and medicine, or about how the world really works: nanobots would take care of it all!

4 Julian Simon from the University of Maryland was one of the most influential cornucopians of the last two decades of the 20th century. His most cited works are: *The Ultimate Resource* (Princeton, NJ: Princeton University Press, 1981) and J. L. Simon and H. Kahn, *The Resourceful Earth* (Oxford: Basil Blackwell, 1984).

5 Electric cars: Bloomberg NEF, *Electric Vehicle Outlook 2019*, https://about.bnef.com/electric-vehicle-outlook/#toc-download. EU carbon: EU, "2050 long-term strategy," https://ec.europa.eu/clima/policies/strategies/2050_en. Global information in 2025: D. Reinsel et al., *The Digitization of the World From Edge to Core* (November 2018), https://www.seagate.com/files/www-content/our-story/trends/files/idc-seagate-dataage-whitepaper.pdf. Global flying in 2037: "IATA Forecast Predicts 8.2 billion Air Travelers in 2037" (October 2018), https://www.iata.org/en/pressroom/pr/2018-10-24-02/.

6 See national long-term fertility trajectories at the World Bank's Data Bank: https://data.worldbank.org/indicator/SP.DYN.TFRT.IN.

7 United Nations, *World Population Prospects 2019*, https://population.un.org/wpp/Download/Standard/Population/.

8 Electric vehicles attracted enormous attention, as well as a great deal of vastly exaggerated expectations, during the second decade of the 21st century. In 2017 one could even read this in *Financial*

Post: "All fossil-fuel vehicles will vanish in 8 years in twin 'death spiral' for big oil and big autos, says study that's shocking the industries." What should have been shocking was the utter lack of technical understanding leading to that ridiculous claim. With about 1.2 billion internal combustion cars on the road in early 2020, this would be some vanishing act during the next five years!

9 It is still unclear when battery electric and conventional vehicles will reach lifetime cost parity, but even when they do, some buyers might still value upfront cost more than any future savings: MIT Energy Initiative, *Insights into Future Mobility* (Cambridge, MA: MIT Energy Initiative, 2019), http://energy.mit.edu/insightsintofuturemobility.

10 For recent sales and long-term forecasts of electric cars adoption, see Insideevs, https://insideevs.com/news/343998/monthly-plug-in-ev-sales-scorecard/; J.P. Morgan Asset Management, *Energy Outlook 2018: Pascal's Wager* (New York: J.P. Morgan, 2018), pp. 10–15.

11 Bloomberg NEF, *Electric Vehicle Outlook 2019*.

12 Michel de Nostredame published his prophecies in 1555, and the true believers have been reading and interpreting them ever since. As for the format, they now have choices ranging from expensive bound facsimiles to Kindle copies.

13 H. Von Foerster et al., "Doomsday: Friday, 13 November, A.D. 2026," *Science* 132 (1960), pp. 1291–1295.

14 P. Ehrlich, *The Population Bomb* (New York: Ballantine Books, 1968), p. xi; R. L. Heilbroner, *An Inquiry into the Human Prospect* (New York: W. W. Norton, 1975), p. 154.

15 Calculated from data in United Nations, *World Population Prospects 2019*.

16 Assuming the median projection from United Nations, *World Population Prospects 2019*.

17 V. Smil, "Peak oil: A catastrophist cult and complex realities," *World Watch* 19 (2006), pp. 22–24; V. Smil, "Peak oil: A retrospective," *IEES Spectrum* (May 2020), pp. 202–221.

18 R. C. Duncan, "The Olduvai theory: Sliding towards the post-industrial age" (1996), http://dieoff. org/page125.

19 For data on undernourishment, see FAO's annual reports. The latest version is: *The State of Food Security and Nutrition*, http://

www.fao.org/3/ca5162en/ca5162en.pdf. For food supplies see: http://www.fao.org/faostat/en/#data/FBS.

20 Calculated from http://www.fao.org/faostat/en/#data/.

21 Data from British Petroleum, *Statistical Review of World Energy*.

22 Data from S. Krikorian, "Preliminary nuclear power facts and figures for 2019," International Atomic Energy Agency (January 2020), https://www.iaea.org/newscenter/news/preliminary-nuclear-power-facts-and-figures-for-2019.

23 M. B. Schiffer, *Spectacular Flops: Game-Changing Technologies That Failed* (Clinton Corners, NY: Eliot Werner Publications, 2019), pp. 157–175.

24 S. Kaufman, *Project Plowshare: The Peaceful Use of Nuclear Explosives in Cold War America* (Ithaca, NY: Cornell University Press, 2013); A. C. Noble, "The Wagon Wheel Project," WyoHistory (November 2014), http://www.wyohistory.org/essays/wagon-wheel-project.

25 On the shrinking climate niche: C. Xu et al., "Future of the human climate niche," *Proceedings of the National Academy of Sciences* 117/21 (2010), pp. 11350–11355. Migrations: A. Lustgarten, "How climate migration will reshape America," *The New York Times* (December 20, 2020). Falling income: M. Burke et al., "Global non-linear effect of temperature on economic production," *Nature* 527 (2015), pp. 235–239. Thunberg prophecy: A. Doyle, "Thunberg says only 'eight years left' to avert 1.5°C warming," Climate Change News (January 2020), https://www.climatechangenews.com/2020/01/21/thunberg-says-eight-years-left-avert-1-5c-warming/.

26 This predilection for catastrophic prophecies is perhaps best explained by human negativity bias: D. Kahneman, *Thinking Fast and Slow* (New York: Farrar, Straus and Giroux, 2011); United Nations, "Only 11 years left to prevent irreversible damage from climate change, speakers warn during General Assembly high-level meeting" (March 2019), https://www.un.org/press/en/2019/ga12131.doc.htm; P. J. Spielmann, "U.N. predicts disaster if global warming not checked," AP News (June 1989), https://apnews.com/bd45c372caf118ec99964ea547880cd0.

27 FII Institute, *A Sustainable Future is Within Our Grasp*, https://fii-institute.org/en/downloads/FIII_Impact_Sustainability_2020.pdf;

J. M. Greer, *Apocalypse Not!* (Hoboken, NJ: Viva Editions, 2011); M. Shellenberger, *Apocalypse Never: Why Environmental Alarmism Hurts Us All* (New York: Harper, 2020).

28 V. Smil, "Perils of long-range energy forecasting: Reflections on looking far ahead," *Technological Forecasting and Social Change* 65 (2000), pp. 251–264.

29 Food and Agriculture Organization, *Yield Gap Analysis of Field Crops: Methods and Case Studies* (Rome: FAO, 2015).

30 Water makes up more than 95 percent of their tissues and they contain no, or negligible, amounts of the two essential macronutrients: dietary proteins and lipids.

31 Its material (steel, plastics, glass) and energy (heating, lighting, air conditioning) costs would be truly astronomical.

32 For energy costs of materials, see Smil, *Making the Modern World*. For minimum energy costs of steel, see J. R. Fruehan et al., *Theoretical Minimum Energies to Produce Steel for Selected Conditions* (Columbia, MD: Energetics, 2000).

33 FAO, "Fertilizers by nutrient" (accessed 2020), http://www.fao.org/faostat/en/#data/RFN.

34 Data from Smil, *Energy Transitions*.

35 Calculated from data in British Petroleum, *Statistical Review of World Energy*.

36 For a glimpse of fascinating discussions of category mistakes, see O. Magidor, *Category Mistakes* (Oxford: Oxford University Press, 2013); W. Kastainer, "Genealogy of a category mistake: A critical intellectual history of the cultural trauma metaphor," *Rethinking History* 8 (2004), pp. 193–221.

37 For the origins of these fundamental inventions, see Smil, *Transforming the Twentieth Century*.

38 Smil, *Prime Movers of Globalization*.

39 A. Engler, "A guide to healthy skepticism of artificial intelligence and coronavirus" (Washington, DC: Brookings Institution, 2020).

40 "CRISPR: Your guide to the gene editing revolution," *New Scientist*, https://www.newscientist.com/round-up/crispr-gene-editing/.

41 Y. N. Harari, *Homo Deus* (New York: Harper, 2018); D. Berlinski, "Godzooks," *Inference* 3/4 (February 2018).

42 E. Trognotti, "Lessons from the history of quarantine, from plague to influenza," *Emerging Infectious Diseases* 19 (2013), pp. 254–259.
43 S. Crawford, "The Next Generation of Wireless—'5G'—Is All Hype," *Wired* (August 2016), https://www.wired.com/2016/08/the-next-generation-of-wireless-5g-is-all-hype/.
44 "Lack of medical supplies 'a national shame,'" BBC News (March 2020); L. Lee and K. N. Das, "Virus fight at risk as world's medical glove capital struggles with lockdown," Reuters (March 2020); L. Peek, "Trump must cut our dependence on Chinese drugs—whatever it takes," *The Hill* (March 2020).
45 The final cost of the 2020 pandemic will not be known for years, but there is no doubt about its order of magnitude: many trillions of dollars. In 2019, the global economic product was close to $90 trillion, and hence it takes a decline of just a few percent to push the cost into the trillions.
46 But we cannot make any final judgment until we get an eventual retrospective worldwide assessment of the pandemic's toll.
47 J. K. Taubenberger et al., "The 1918 influenza pandemic: 100 years of questions answered and unanswered," *Science Translational Medicine* 11/502 (July 2019), eaau5485; Morens et al., "Predominant role of bacterial pneumonia as a cause of death in pandemic influenza: Implications for pandemic influenza preparedness," *Journal of Infectious Disease* 198 (2008), pp. 962–970.
48 "The 2008 financial crisis explained," History Extra (2020), https://www.historyextra.com/period/modern/financial-crisis-crash-explained-facts-causes/.
49 The largest cruise ships now accommodate more than 6,000 passengers; crew adds up an additional 30–35 percent. Marine Insight, "Top 10 Largest Cruise Ships in 2020," https://www.marineinsight.com/know-more/top-10-largest-cruise-ships-2017/.
50 R. L. Zijdeman and F. R. de Silva, "Life expectancy since 1820," in J. L. van Zanden et al., eds., *How Was Life? Global Well-Being since 1820* (Paris: OECD, 2014), pp. 101–116.
51 These excess mortalities can be seen on regularly updated websites by the European Mortality Monitoring (https://www.euromomo.eu/) for EU countries, and by the Centers for Disease

Control (https://www.cdc.gov/nchs/nvss/vsrr/covid19/excess_deaths.htm) for the US.

52 Detailed age-specific population projections for all of the world's countries and regions are available at: https://population.un.org/wpp/Download/Standard/Population/.

53 American Cancer Society, "Survival Rates for Childhood Leukemias,"https://www.cancer.org/cancer/leukemia-in-children/detection-diagnosis-staging/survival-rates.html.

54 US Department of Defense, *Narrative Summaries of Accidents Involving U.S. Nuclear Weapons 1950–1980* (1980), https://nsarchive.files.wordpress.com/2010/04/635.pdf; S. Shuster, "Stanislav Petrov, the Russian officer who averted a nuclear war, feared history repeating itself," *Time* (September 19, 2017).

55 The most detailed report of the disaster (including five technical volumes) is: International Atomic Energy Agency, *The Fukushima Daiichi Accident* (Vienna: IAEA, 2015). The National Diet of Japan issued its official report: *The Official Report of the Fukushima Nuclear Accident Independent Investigation Commission*, https://www.nirs.org/wp-content/uploads/fukushima/naiic_report.pdf.

56 For Boeing's official announcements, see 737 MAX updates at https://www.boeing.com/737-max-updates/en-ca/737 MAX. For critical appraisals see, among many others: D. Campbell, "Redline," *The Verge* (May 2019); D. Campbell, "The ancient computers on Boeing 737 MAX are holding up a fix," *The Verge* (April 2020).

57 In 2018, the shares of global CO_2 emissions were as follows: the top emitter (China) very close to 30 percent; the top two (China and the US) a bit over 43 percent; the top five (China, USA, India, Russia, Japan) 51 percent; top 10 (add Germany, Iran, South Korea, Saudi Arabia, and Canada) almost exactly two-thirds: Olivier and Peters, *Global CO_2 emissions from fossil fuel use and cement production per country, 1970–2018*.

58 This necessity for very long-term commitment further diminishes the likelihood that such disparate actors as China and the US or India and Saudi Arabia will agree on a generally acceptable and durable way to proceed.

59 Ramsey's classic appraisal is unequivocal: "It is assumed that we do not discount later enjoyments in comparison with earlier ones, a practice which is ethically indefensible and arises merely from the weakness of imagination." F. P. Ramsey, "A mathematical theory of saving," *The Economic Journal* 38 (1928), p. 543. Of course, such an unyielding position is quite impractical.

60 C. Tebaldi and P. Friedlingstein, "Delayed detection of climate mitigation benefits due to climate inertia and variability," *Proceedings of the National Academy of Sciences* 110 (2013), pp. 17229–17234; J. Marotzke, "Quantifying the irreducible uncertainty in near-term climate projections," *Wiley Interdisciplinary Review: Climate Change* 10 (2018), pp. 1–12; B. H. Samset et al., "Delayed emergence of a global temperature response after emission mitigation," *Nature Communications* 11 (2020), article 3261.

61 P. T. Brown et al., "Break-even year: a concept for understanding intergenerational trade-offs in climate change mitigation policy," *Environmental Research Communications* 2 (2020), 095002. Using the same model, Ken Caldeira calculated the internal rate of return on the abatement investment ramping (as stated by many recent national goals) to zero carbon by 2050 and the starting date of the positive return (when avoided climate damage exceeds abatement expense): the rate is about 2.7 percent and the positive return does not come until early next century.

62 High forecast: United Nations, *World Population Prospects 2019*. Low forecast: S. E. Vollset et al., "Fertility, mortality, migration, and population scenarios for 195 countries and territories from 2017 to 2100: a forecasting analysis for the Global Burden of Disease Study," *The Lancet* (July 14, 2020).

Appendix: Understanding Numbers: Orders of Magnitude

1 M. M. M. Mazzocco et al., "Preschoolers' precision of the approximate number system predicts later school mathematics performance," *PLoS ONE* 6/9 (2011), e23749.

2 United States Census, *HINC-01. Selected Characteristics of Households by Total Money Income* (2019), https://www.census.gov/data/tables/time-series/demo/income-poverty/cps-hinc/hinc-01.html; Credit Suisse, *Global Wealth Report* (2019), https://www.credit-suisse.com/about-us/en/reports-research/global-wealth-report.html; J. Ponciano, "Winners/Losers: The world's 25 richest billionaires have gained nearly $255 billion in just two months," *Forbes* (May 23, 2020).

3 V. Smil, "Animals vs. artifacts: Which are more diverse?" *Spectrum IEEE* (August 2019), p. 21.

4 The power of prime movers is reviewed in V. Smil, *Energy in Civilization: A History* (Cambridge, MA: MIT Press, 2017), pp. 130–146.

Acknowledgments

Thanks to Connor Brown, my London editor, for giving me another chance to write a wide-ranging book, and to my son David (at the Ontario Institute for Cancer Research) for being its first reader and critic.

Index